INDUSTRIAL SECURITY

Industrial Security

A. E. Warne, FIISec
D. G. Brown, BA, FIISec

Barry Rose (Publishers) Ltd
Chichester and London
1977

ISBN 0 85992 090 9
Published by Barry Rose (Publishers) Limited
Chichester and London

Printed in England by
Unwin Brothers Limited
Old Woking, Surrey, England

Foreword

We set out to produce a book which would be a training aid for the middle and lower ranks of the security profession but from our deep interest in, and enthusiasm for, security education we soon realized that such a training manual would lose much of its meaning if it did not include a detailed discussion of company policy which, in every business concern, must form the source from which all security action stems. The book now offers to management in industry and commerce practical advice and guidance on many aspects of company policy affecting good security practice.

We believe that there is good purpose in presenting a single volume to both management and to the security practitioner since, for maximum effectiveness, it is essential for security personnel at all levels to know and understand the problems associated with management policies as they affect security and for management to know of the complexities inherent in an adequate performance of the security function.

We have incorporated into the book firm statements of law coupled with practical interpretation of that law as, in our views, it affects the security of the business world. Security problems are discussed from differing viewpoints which we trust will enable the reader to formulate a balanced opinion. There are no fanciful considerations of highly improbable situations and the reader's time is not absorbed by technical detail of interest only to academic theorists.

From a welter of new law to be entered upon the statute book over recent years we have selected such matters as the Code of Practice Relating to Discipline Practice and Procedure arising from the Employment Protection Act 1975 to which we have added some comment upon the practical operation of the Code in the interests of good industrial relations; the Rehabilitation of Offenders Act 1975 and the limitations which that Act imposes upon inquiries when recruiting new staff; the Health and Safety at Work etc Act 1974 with particular regard for a contemporary approach to company safety policy; the Fire Precautions Act 1971 and the extension of the fire certificate requirements of that Act to factories, offices, shops and railway premises effective from January 1, 1977.

We have considered the responsibilities of security management and, as originally intended, we provide information for training at all levels of the security profession from the routine of physical patrolling to the skills of the investigator. From the standpoint of experienced security practitioners we discuss such contentious items in company policy as the searching of personnel, their vehicles and their tool and clothes lockers; company attitudes towards employees and others involved in criminal acts against the company, against their colleagues or against society beyond the bounds of their employment; and the quantity and quality of the protection to be afforded to premises. We have taken a careful look at the responsibility of middle management for good industrial relations; at contingency planning to meet the aftermath of catastrophe, fire, or other emergency situations; and at current threats to industry and commerce from the bomber, the saboteur, the arsonist and the industrial spy.

Introductory paragraphs explain the intended purpose and use which can be made of each of the four Parts of the book.

We have not found the space to enlarge upon the equipment side of the security industry nor to dwell upon security duty performed in specialized environments such as seaports, airports or overseas locations nor where the work involves special skills required for the protection of VIP visitors, of company officials travelling abroad, of works of art in transit or on exhibition, or in the routine matters of hotel security, or problems associated with the retail trade. We have aimed at the more general field as a compendium of studies to enlarge and enrich an understanding of industrial security for the widest possible readership whose interests may lie in a study of social sciences; as a manager in a business house trying to formulate a security policy which will be both acceptable and workable; as a security administrator; or as a fledgling security operative. For all those readers we have attempted to supplement other text books rather than to repeat or replace them.

We greatly appreciate the advice, assistance and encouragement which we have received from many of our colleagues who believe, as we do, that the advance of knowledge in the security arena requires the printed word. It is toward the improvement of the theory and the practice of industrial security that this book is directed and we trust that we have included in it sufficient material to be of interest and benefit to management as well as to our colleagues at all levels of the security profession.

ALBERT E WARNE
DONALD G BROWN

Contents

PART ONE

Security is a Management Responsibility

Introduction

A student once wrote in an essay that the purpose of security in the business world is to enhance the profitability of companies by providing safeguards against losses caused by theft, fire, fraud, waste, damage and trespass. In a nutshell that student had described precisely how and why security departments could be justified and paid for – he had identified the task of management which is to increase profitability, with the preventive and protective elements of good security – and he thereby established that security is a management responsibility.

Not only is it a responsibility of the board room, of corporations and councils and of top executive management but with each delegation of responsibility below those levels there must go a slice of responsibility for security within the delegated domain – and this trend continues through senior and middle management to first line managers, supervisors, foremen and overseers. Each in his own sphere of activity can enhance the profitability of the concern by applying security standards and each should have the burden of applying those standards written into his job description as a reminder that he is being remunerated for that aspect of his management role.

The appointment of a full time security adviser, consultant or manager and the setting up of a security department in no way relieves management at any level of its obligations to prevent theft, fire, fraud, waste, damage or trespass and to bring to justice those who offend against the law or company rules or who by any misconduct or breach of security otherwise reduce profitability.

The security chief and his department are there to assist management just as a police force and public fire brigade are there to support society in its tasks of preserving the peace and preventing catastrophe.

In assessing the security standards which have to be maintained every level of management is entitled to look upwards for guidance and therefore those standards must be determined at the top by the Board, by the Corporation, by the Council and by the Top Executive. Company Security Policy, as we will choose to call it, must be a settled and decided policy and must be given publicity within the concern so that all employees and others having contact with the business may know the company's intentions and basis for its decisions when individual cases are considered.

Factors which make up Company Security Policy must include its attitudes towards

- i those who break the law or rules of conduct during their employment or contact with the business;
- ii a condition of employment that requires employees to accept a policy of personal search;
- iii the vetting and character enquiry for new applicants for employment;
- iv management involvement in security matters at all levels and its support for a security department;
- v the volume and extent of physical protection to be given to buildings and contents or protection by electronic and other sophisticated detectors of intruders and for fire;
- vi the manner by which it is proposed to guard against leakage and loss of confidential information;
- vii the arrangements it will make for emergency planning to safeguard the health, safety and welfare of employees and to ensure a continuous place of work.

Just as there are many factors affecting company policy so are there very many more considerations affecting each of those factors and the purpose of Part I of this book is firstly to consider the short history of industrial security in order to more fully appreciate and understand the contemporary scene and then to examine in some detail the factors which make up Company Security Policy.

CHAPTER ONE

Twenty-five Years of Change

Until the middle of the twentieth century security in industry and commerce was the ugly duckling in those few companies which gave it recognition – it was then under sufference with no legal standing or accepted status. Today, 25 years later, it is a lusty young professional making its presence known and respected in the tumultuous world of trade, in government departments, in nationalized industries and in public undertakings. That quarter century ago security was composed of gatekeepers and nightwatchmen, faithful company servants serving out their pre-retirement years – endeavouring to stem the rising flood of crime that in post war years had been sweeping the western world.

Burglar alarms had been known for more than 30 years but only a few shopkeepers, art galleries, spirit and tobacco warehouses and clothing manufacturers installed primitive systems in attempts to thwart thieves. In the 'fifties burglar alarms hit the headlines by the number of arrests effected with the aid of the new silent 999 call to police but countermeasures were quickly taken by thieves who stole radio equipment with which to monitor police communications to know precisely when to expect the arrival of the law. Those were indeed changing days. The mood of the western world was changing and crime and criminals were changing with it – to throw off discriplines, restraints and controls, to taste the freedom of self expression safely executed amongst the fellowship of the gang, and to abandon respect for law and order and for the rights and privileges of others. The day had arrived of the pay snatch bandit – of armed robbery from banks and post offices – of the use of firearms in the commission of all kinds of theft – of the thermic lance – and of the criminal's accomplice willing to be dumped in Epping Forest bound hand and foot or bludgeoned unmercifully to make the loss of his shop or vehicle keys look good.

It is said that much of that mood had its origins in wartime conditions when hitherto honest people savoured the fruits of black market dealings with goods in short supply; in the ease with which loot from occupied or re-occupied lands passed into the hands of the occupying forces; in the freedom of trade concerning surplus war materials, much of which was in no way surplus at the time of the trading; and in post-war years of rationing and shortages when graft was well nigh essential to woo what small amount of business was available. Getting the most out of winning the peace – of easy pickings, of something for nothing, of one-up-manship in possessing something not available to one's neighbours became an established way of life for large sections of the populace including their growing children. It was almost the thing to do to defraud or to steal from one's employer or one's school, to take without asking and as a right, to shoplift, to idle or skive for one's personal interests at work and to commit spiteful damage in retaliation for any attempt to impose restraint or control. The anti-authoritarian code of living grew strong from the burden of increasing taxation, an inevitable aftermath of war, which created in people an evasiveness and penchant for secretive transactions; from improving standards of living after years of mediocrity which nurtured greed and avarice; from the arrival of the permissive society with its lowering of personal disciplines and moral standards; from the one class society outlook with its freedom of expression to speak out without reserve regardless of hurt feelings or offence to good taste; from the lowering of severity by the Courts; from the growth industry of looking after the underdog and from the flood of welfare state handouts.

Crime figures known to the authorities doubled those of pre-war years and in some classes of crime the figures were quadrupled. Police forces, struggling to regain or maintain pre-war manpower strengths and stretched beyond the reasonable calls of duty, frequently found frustration in the reluctance by the Courts to impose penalties commensurate with this new gravity of crime. No longer could property be left unattended in the street, car park or public rooms without taking security measures to ensure its presence there upon the return of the lawful owner or user; no longer could a manufacturer despatch a laden lorry without taking a series of protective measures to ensure delivery intact at destination; no longer could administrative staff be given one hundred per cent trust for accounting systems and financial transactions without frequent internal audits; and no longer could banks and safe deposits be regarded as impregnable unassailable strongholds against robbers and thieves.

Such was the sorry situation that existed in the 'fifties, worsened through the 'sixties and continued into the 'seventies and such is the sorry situation that has caused the growth of the security weakling into the lusty young professional about whom this chapter commenced. Industry and commerce, government departments, national and public departments, mines, docks, airports and railways have seen that authority in the person of the Police cannot be relied upon to contain the modern mood for crime and cannot deal with the thousands of criminal activities that have become the bane of the business world. Home Office and Police Crime Prevention Departments now encourage an attitude of self preservation and companies and industries have taken their own measures to protect their assets and the personal possessions of their employees at work, and to identify and bring to justice those who are endeavouring to live by crime.

Crime prevention engineers and electronic experts have shown immense originality and resource by producing sophisticated intruder detector systems, in designing secure cash carrying vehicles and apparatus and in formulating personal identification systems, in inventing stronger thief-proof locks, safes and similar hardware. Photographic and television experts have focused attention upon their specialized equipment for the prevention and detection of crime. Service companies offering the loan or hire of equipment, manpower, dogs and transport have found a sellers' market awaiting their rapid expansion.

Small wonder that there have been rumblings of concern in the corridors of power about the growth and power of private armies and about a disturbing usurpation of the authority of the police forces of the country. Those engaged within the security industry know that these forebodings are unwarranted and that there is no attempt or intention by the security industry to detract from the efficiency and expertise of the police but they also know that there is a limit to the ability of police and the courts to cope with the overall confrontation of current crime. Experience gained in the fields of prevention and detection of crime has shown private and public business concerns that the private citizen still has rights, powers and duties within the law for coping with criminal situations and such concerns are becoming increasingly prepared to take whatever action is lawful and for this purpose they create security departments.

Security departments are now essential units of most businesses, of as much importance as buying and sales divisions, of production and transport departments and of statistical and finance branches. Security departments are recognized by workers representatives in their bargaining with employers and are accepted by the rank and file so long as security activity does not interfere with their conditions of employment nor with the liberties which they take with their employer's time.

Today the security industry is made up of:

i The security departments of private and nationalised industry, of government departments, docks, railways, airports and transport systems generally, banking and of all manufacturing, wholesale, retail and distributive trades. Security is one factor common to all and yet is one of the least talked about or recognized facets of direct employment.

ii Security service companies which provide similar services to those of security departments but on a contract hire basis to supply manpower and equipment as and when required for the protection of premises and property; for handling cash in transit, or to conduct investigations into alleged crime.

- iii Security consultants and advisers working alone or in partnerships and offering professional advice for professional fees.
- iv The manufacturers, wholesalers, distributors and installers of security equipment ranging widely from safes, locks, gates, grilles, barbed wire and barriers through special lighting, alarm systems, CCTV, special glass, highly sensitive smoke, heat and movement detectors, fire quenchers and other sophisticated mechanical electric and electronic devices for preventing and detecting intrusion of premises, interference with property and incipient fire.

Any board of directors or principal of a business who is not taking preventive and protective action under one or other of those branches of the industry is not in touch with the realities of the contemporary scene and is most likely to be taking far greater risks with the profitability of his business than is reasonable or else he is grossly over-insured.

The security industry is one of the most progressive industries demanding high standards of skill and professionalism with continual updating, improvement and training for all ranks and at all levels. There are no reliable statistics available to accurately assess the number of persons employed in the industry but an average taken from the informed guesses known to the authors is that 150,000 in Great Britain would be about the correct figure – and of course industrial security is now world wide.

The growth of the industry has been in step with the growth of crime and it is only fair to say that history seems to be repeating itself because the late Sir Richard Jackson, CBE, a former Assistant Commissioner of the Metropolitan Police commented in his book *Occupied with Crime* – 'In 1751 Henry Fielding one of the Bow Street magistrates wrote in a treatise called "An enquiry into the cause of the late increase in robbers" that growth of crime was a by-product of growing public wealth and it was his experience that most thieves stole luxuries rather than necessities.' Isn't that just what this Chapter has been saying – the more luxuries there are available the more will there be thieves trying to get hold of them for nought. Insofar as these islands are concerned the expansion and experience of industrial and commercial security may not have eliminated crime but there is little doubt that it has considerably assisted the law enforcement agency to contain it.

CHAPTER TWO

Company Security Policy

Prosecution

The Introduction to Part I refers to a company's attitude towards those who break the rules of law and conduct during their employment or when in contact with an employer. Company policy should clearly express the company's attitude and intentions towards

i alleged or disclosed crimes against the company and its property (a) by company employees, or (b) by others;

ii alleged or disclosed crimes against the person or property of em-employees or others on company premises (a) by other company employees, or (b) by others;

iii employees convicted of crime outside the orbit of their employment.

Opinion amongst top management in commerce and industry and amongst their senior security officers as to the lines of action that a security policy should specify may be widely divergent and it is for each board of directors or management to make its own policy decisions based upon certain broad outlines and considerations.

Whatever diversity of opinion may have existed in the past it is very evident that existing policies should have been founded upon the basis of paragraphs 130 to 133 of the Industrial Relations Code of Practice published By Authority in 1972 in compliance with Section 2 Industrial Relations Act 1971 and re-enacted without amendment by the Trade Unions and Labour Relations Act 1974 (First Schedule). Guidance for future policies will be found in the *Draft Code of Practice relating to Disciplinary Practice and Procedure* published by the Advisory Conciliation and Arbitration Service (ACAS) which has been set up in pursuance of Section 1 of the Employment Protection Act 1975.

The ACAS has issued its draft Code of Practice on discipline by authority of Section 6(1) of the 1975 Act and this draft closely follows paragraphs 130–133 of the 1972 Code which it is intended to replace. The new Code may well become part of the law by the time this book is published. Accordingly readers should recognize the following as a draft recommendation being considered by ACAS at the time of going to print and should compare it for any later modifications or changes in the text with the Code as finally approved by Parliament.

Draft Code of Practice on Disciplinary Practice and Procedures

Introduction

1 This document seeks to give practical guidance on how to draw up disciplinary rules and procedures and how to operate them effectively. Its aim is to help employers and trade unions as well as individual employees wherever they are employed and regardless of the size of the organisation in which they work.

Why have disciplinary rules and procedures?

2 Disciplinary rules and procedures help promote fairness and order in the treatment of individuals and in the conduct of industrial relations. Rules set standards of performance and behaviour at work; procedure helps to ensure that the standards are adhered to and also provides a fair method of dealing with any alleged failure to observe them.

3 It is important that employees know what standards of performance and behaviour are expected of them and the Contracts of Employment Act 1972 (as amended by the Employment Protection Act 1975) requires employers to provide written information for their employees about certain aspects of their disciplinary rules and procedures.

4 The importance of disciplinary rules and procedures has also been recognised by the law relating to dismissals, since the grounds for dismissal and the way in which the dismissal has been handled can be challenged before an industrial tribunal. Where either of these is shown to have been unfair the employer may become liable to pay considerable sums of compensation to the employees concerned.

Note: Employment Protection Act 1975, section 71–76 inclusive. A tribunal can order the payment of: a basic award (maximum £2,400), a compensatory award (maximum £5,200), and an additional award if an order for re-instatement or re-engagement is not complied with (maximum £4,160).

Formulating the policy

5 Management is responsible for ensuring that there are adequate disciplinary rules and procedures within their organisation and the initiative for establishing these will normally lie with them. However, if they are to be effective the rules and procedures need to be acceptable to the employees who are to be covered and to those who operate them. The aim should therefore be to secure the maximum involvement of employees and all levels of management in their formulation.

6 It is desirable that trade union officials should participate fully with management in formulating and agreeing those procedural arrangements

and disciplinary rules which will apply to their members. This would ensure that the disciplinary policy is more acceptable and would give trade union officials greater opportunity to see that it is operated consistently and fairly.

Note: Throughout this Code, trade union official is defined as in section 30(1) of the Trade Union and Labour Relations Act 1974 and where appropriate includes safety representatives appointed under section 2(4) of the Health and Safety at Work etc Act 1974.

The basic rules

7 The basic rules should be those which are necessary for the efficient and safe performance of work and for the maintenance of satisfactory relations within the workforce and between employees and management. They will vary according to particular circumstances such as the type of work, working conditions and size of establishment. Some organisations will need specific rules for safety, some for hygiene, whilst in others the essential continuity of work will make time-keeping and attendance important considerations; sometimes particular jobs will require special rules. In every case unnecessary rules should be avoided.

Note: It should be noted that this Code is not concerned with the duties imposed on employers by the Health & Safety at Work etc Act 1974 to provide employees with information about rules and procedures concerned with health and safety at work.

8 If rules are reasonable they will be better respected by the employees concerned. However, in their preparation a balance needs to be drawn between clarity (in order not to render the rules meaningless) and flexibility (in order to leave room for mitigating circumstances).

9 Rules should be readily available. One way of achieving this is to post them on notice boards at the employee's place of work but it will often be better practice to give a copy to each individual employee. Many employers include them in a handbook which is issued to their employees.

10 Employees should know the likely consequences of breaking rules and they should be left in no doubt as to the type of misconduct which is serious enough to warrant suspension pending investigation or even instant dismissal. In practice, rules of which a breach is serious enough to warrant suspension or dismissal should be separated from those which are more appropriately dealt with through a procedure which gives an opportunity for standards of conduct to be improved.

Essential features of disciplinary procedures

11 Disciplinary procedures should:

a) Provide for speedy operation.

b) Be in writing.
c) Specify to whom they apply.
d) State the range of disciplinary actions which can be taken.
e) Specify who has the authority to take the various forms of disciplinary action, ensuring that immediate superiors do not normally have the power to dismiss without reference to more senior management.
f) Provide for the individual to be informed of the complaint against him and give him the opportunity to state his case before a decision is reached.
g) Give the individual the right to be accompanied by a representative of his choice at any stage.
h) Provide for a written warning or penalty to be acknowledged in writing by the individual or for the written communication to be handed to him in the presence of his representative.
i) Ensure that, except for serious misconduct, no employee is dismissed for a first breach of discipline.
j) Provide for suspension to ensure that, except in special circumstances, no dismissal takes effect until the case has been carefully investigated.
k) Provide a right of appeal to some level of management, joint committee or independent third party, not previously directly involved in the case.
l) Ensure that the individual is given the reason for any penalty imposed.
m) Ensure that the individual is informed of his right of appeal.

The procedure in operation

12 When a disciplinary matter arises, the supervisor or manager should first establish the facts promptly before recollections fade, taking into account the statements of any available witnesses. Before making a decision or imposing a penalty he should interview the individual giving him the opportunity to state his case and advising him of his rights under the procedure, including the right to be accompanied by a representative of his choice.

13 Where recognised standards of performance or behaviour have not been observed, the action taken will depend on the circumstances. In those cases of alleged serious misconduct for which the rules provide for summary dismissal, the first step should normally be suspension whilst the case is investigated. During this process the employee should be entitled to his rights under the procedure. In all other cases the procedure would normally be as follows:

a) In the case of minor offences the immediate supervisor should give an oral warning for the purpose of improving future performance.

b) If further action proved necessary, or if the issue were more serious, there should be a written warning setting out the circumstances.
c) Further misconduct might warrant a final written warning which should contain a statement that any recurrence would lead to suspension or dismissal or some other penalty, as the case may be.
d) The final step might be disciplinary transfer, or disciplinary suspension without pay (but only if these are allowed for by an express or implied condition of the contract of employment), or dismissal, according to the nature of the misconduct. Suspension without pay should not be used needlessly and should not normally be for a prolonged period.

14 Details of any disciplinary action should be given in writing to the employee and, if he so wishes, to his representative. At the same time the employee should be told of his right of appeal, how to make it and to whom.

15 When determining the disciplinary action to be taken the supervisor or manager should bear in mind the need to satisfy the test of reasonableness in all the circumstances. He should, so far as he is able, take account of any mitigating factors.

16 Special consideration should be given to the way in which disciplinary procedures are to operate in exceptional cases. For example:

a) *Employees to whom the full procedure is not immediately available* Special provisions should be made for the handling of disciplinary matters amongst nightshift workers, workers in isolated depots and others who may pose particular problems because no one is present with the necessary authority to take disciplinary action.
b) *Trade union officials* Disciplinary action against a trade union official can lead to a serious dispute if it is seen as an attack on the union's functions. Although an oral warning might be given, no other action should be taken until the circumstances of the case have been discussed with a senior trade union representative or full-time official.
c) *Criminal offences outside employment* These should not be treated as automatic reasons for dismissal, regardless of whether the offence has any relevance to the duties of the individual as an employee. The main considerations should be whether the offence is one that makes the individual unsuitable for his type of work or unacceptable to other employees, and if so, whether alternative jobs are available which would be suitable.
   Employees should not normally be dismissed simply because a charge against them is pending or because they are absent through having been remanded in custody.

d) *New employees working a probationary period with the employer* It may not be appropriate for employees on probation to have access to the full disciplinary procedure when they are demoted or dismissed for failure to meet the required standards. They should, however, at least be given warning in a formal manner so that they may understand the standards required of them.

Appeals

17 Grievance procedures are sometimes used for dealing with disciplinary appeals though it is normally more appropriate to keep the two kinds of procedure separate. Disciplinary issues are essentially matters confined within the organisation and need to be dealt with speedily. The external stages of a grievance procedure may however be the appropriate machinery for dealing with appeals against disciplinary action where a final decision within the organization is contested or where the matter becomes a collective issue between management and a trade union.

18 Independent arbitration is sometimes an appropriate means of resolving disciplinary issues. Where the parties concerned agree, it may constitute the final stage of procedure.

Records

19 Records should be kept, detailing the nature of any breach of disciplinary rules, the action taken and the reasons for it, whether an appeal was lodged, its outcome and any subsequent developments. These records should be carefully safeguarded and kept confidential.

20 Except in special circumstances breaches of disciplinary rules should be removed from an employee's record after an appropriate period of satisfactory conduct.

Further action

21 Rules and procedures should be reviewed periodically in the light of any developments in employment legislation or industrial relations practice and, if necessary, revised in order to ensure their continuing relevance and effectiveness. Such amendments and any additional rules imposing new obligations should only be introduced after full discussion with employees' representatives, reasonable notice has been given, and all employees have been informed.

Formulating a Policy

In the past many employers whether in large companies or small have never drawn up, or have never announced to their employees, a policy founded upon a code of discipline. Serious thought should now be given by all employers to these matters in order to keep abreast of changing social attitudes and to protect themselves from the serious financial liabilities which non compliance with the Code can engender.

The point contained in paragraph 15 of the Draft Code about 'reasonableness' should convey to managers and security personnel thoughts of a middle course which regards prosecution through the courts in every case of crime as being too severe and verbal reprimand in every case as being too easy going. The middle course policy considers that each case should be dealt with on its merits.

There is strong support for the view that a quick decision as to how a case is to be handled is essential so that individuals and their representatives are not kept waiting in suspense as to whether or not police will be called in to deal with a matter affecting them. If police are to be called it is important that there should be no loss of time which could mar police investigations. If police are not being called it is fair to inform the individual promptly that that decision has been taken and that he/she will hear the result of the company's deliberations with regard to internal discipline in due course.

It is generally accepted that in the preparation of a policy regard must be given to the alternative possibilities of whether or not company property is involved and whether or not the accused is an employee. The rights of a man or woman whose person or property has been violated must be respected as to whether he or she wishes police to be called or will be satisfied if internal disciplinary action is taken against an offender or wishes for no action to be taken at all.

The Legal Position

Persons who have been assaulted or whose property has been criminally damaged or stolen are under no legal obligation or compulsion to prosecute those responsible for that injury, loss or damage.

Section 4(1) Criminal Law Act 1967 enacts – 'Where a person has committed an arrestable offence, any other person who, knowing or believing him to be guilty of the offence or of some other arrestable offence, does without lawful authority or reasonable excuse any act with intent to impede his apprehension or prosecution shall be guilty of an offence.'

Section 5(1) Criminal Law Act 1967 enacts – 'Where a person has committed an arrestable offence, any other person who, knowing or believing that the offence or some other arrestable offence has been committed, and that he has information which might be of material assistance in securing the prosecution or conviction of an offender for it, accepts or agrees to accept for not disclosing that information any consideration other than the making good of loss or compensation for that loss or injury, shall be liable on conviction on indictment to imprisonment for not more than two years.'

(The term arrestable offence includes all serious crime such as theft, other crimes involving theft, assault and criminal damage. See Chapter 19.)

Neither of those two Sections make prosecution obligatory.

Section 4(1) makes it an offence to *do any act* which is intended to impede the arrest or prosecution of a person for committing an arrestable offence and therefore does not relate to *omitting* to take the action of causing the accused to be arrested or prosecuted.

Section 5(1) makes it an offence to take a bribe or gift (which it calls a consideration) for not disclosing information which might lead to a prosecution and conviction of a person for committing an arrestable offence but the Section specifically excludes from the offence the taking back of stolen property and the acceptance by the victim of compensation for loss, damage or injury. What is more the law does not now compel any person to give evidence of what he knows about an alleged offender. If he does so he does it voluntarily and as a duty to society against wrongdoers. An employee who has had personal property stolen which is later recovered from the possession of the thief is in the same position as a company whose property has been stolen and recovered in like circumstances – each loser may make his own decision as to whether or not to prosecute the thief.

Having said all that it cannot be too clearly stated that Her Majesty's Judges when presenting to the Home Office their revised version of The Judges Rules in 1964 reiterated the principle that it is the duty of all citizens to assist the police to discover and to apprehend offenders. It is therefore very much a matter of personal point of view for each victim or witness of crime to be guided by his or her sense of public duty towards the overall preservation of law and order in this realm. Whenever police are called in a policy of frankness and co-operation with them must follow. Furthermore there are some crimes which are of such a serious nature or so detrimental to public order that the police may decide to prosecute even when a victim does not wish to press a charge or to support a prosecution.

Loss of Personal Property

When an employee or other individual on company premises has personal property stolen it is in the loser's interest in most cases for the matter to be reported to the police who should be given facilities by the company for an investigation on the premises and to interview employees or others in the course of that investigation. Generally making it known that police will be called to investigate thefts of personal property should act as a deterrent against the commission of such offences.

Loss of Contractors' Property on Company Premises

Any reported loss of property by a contractor on company premises should be recorded and investigated in the same manner as the loss of company owned property. Additionally the contractor is entitled to call in the police to investigate his loss.

Loss of Company Property

1 Theft by Outsiders

If theft of company property is alleged to have been committed by any person other than an employee or contractor's employee, the police should be called without delay.

2 Theft by Contractors Employees

If the accused is an employee of a contractor, and if police proceedings are not being instituted by the company, the facts should be reported to the contractor. It may be intimated to the contractor that were the accused an employee of the company offended against, the matter would be dealt with in a certain way and it may be stated whether or not the man should continue to visit or work on the premises but no other pressure should be applied to the contractor to deal with the man in any particular way.

3 Theft by Employee

When an employee admits to stealing company property the action taken by the manager or senior security officer who is interviewing the employee is of vital importance

a) in order to be fair to the individual;
b) in order to protect the company's interests; and
c) in order to safeguard any court proceedings that may follow.

The interviewer should forthwith make a written record of the exact words used by the employee in making that confession and if he has not previously cautioned the employee he should forthwith caution him in the words given in The Judges Rules, Rule II as follows – 'You are not obliged to say anything unless you wish to do so but what you say may be put into writing and given in evidence.' The interviewer should record the time at which that caution is given. The interviewer may then ask the employee whether he/she will sign a statement containing the oral information just given to the interviewer. If this is agreed the procedure outlined in Rule IV of The Judges Rules should be followed. (See Chapter 18 page 205.)

The Judges Rules place an obligation upon 'police officers and persons other than police officers charged with the duty of investigating offences or charging offenders' to comply with the Rules and this is not normally regarded to include industrial or commercial managers other than security managers and security investigators but there are two very sound reasons why any manager dealing with this kind of disciplinary matter should comply with the Rules if practicable.

i If it is subsequently decided to prosecute the employee for theft to which charge there is a plea of not guilty the statement of confession may be tendered in support of other evidence and will be good evidence if it has been properly taken under caution.

ii If it is subsequently decided to dismiss the employee for gross misconduct any written statement showing grounds for dismissal supplied to the employee on request and in accordance with Section 70 Employment Protection Act 1975 may contain the comment that he/she confessed to the theft of company property. In the event of any subsequent appeal for unfair dismissal the signed statement properly taken in accordance with The Judges Rules must be valid evidence that the dismissal was fair on the grounds of admitted gross misconduct.

When reaching a decision whether to dismiss the employee or to call in the police with a view to prosecution for an admitted theft the employer should not be influenced so much by sentiment for the accused, nor by an assessment of his worth to the company as a skilled employee, as by consideration of one or more of the following factors:

i whether this loss seriously interferes with customer relations or trading efficiency;

ii whether this is one incident in a series involving property of high value;

iii whether lenient action now is likely to encourage others to commit similar acts;

iv whether there is a possibility that police may discover other stolen property at his home or elsewhere in his possession during the course of their enquiries;

v whether this man's action is linked with that of others who are already being investigated or prosecuted by police;

vi whether the calling of police will seriously affect trading operations for example (1) if the accused is a lorry driver will police investigation delay this driver's departure with goods urgently required by a customer? – is there a replacement driver? – how else can that load be delivered to the customer? – does the gravity of the accusation rule out ideas of entrusting this man with this or any other load? – have the stolen goods been recovered? (2) if the accused is a wharfside crane operator will prosecution result in delay occurring in the arrival or departure of watercraft with which this man's duties are so closely related? – what value can be placed upon such a delay? –

and what alternative arrangements can be made to meet company needs of craft movement in the immediate future?

vii whether there has been an honest and well kept record of annual review of this employee showing good or bad character and whether there have been previous defaults serious or otherwise.

In the event of an accused employee denying that he has stolen company property a point is reached at which management must make a choice between three courses of action.

The first course is to drop the enquiry and to take no further action. This course is unthinkable because (a) it places the company in an extremely weak position for any later disciplinary action against anyone; and (b) it is unfair to the man himself and to other employees in the company to permit him to continue to work should he be accused but not prosecuted and therefore not proven to be guilty or innocent and it leaves the individual under a cloud of suspicion during which he will foster resentment towards his accuser whilst suffering animosity towards himself from other employees thus creating a very unsavoury working 'atmosphere'.

The second course is to dismiss the employee forthwith for gross misconduct. In taking that decision the employer places himself in the position of judge and jury and decides that the accused is guilty against that individual's denial. The employer thereby renders himself liable to be called upon to satisfy an industrial tribunal that the dismissal was not unfair. It may depend upon the strength of the evidence available to back up this dismissal but it is well to remember that very many cases of highly suspicious conduct do not result in convictions at court and therefore a presumption of guilt is a most unreliable decision to take. In any event an employer should not place himself in the position of 'defendant' before a tribunal because he has peremptorily saved the suspect from facing a theft charge in a criminal court. Furthermore if a tribunal finds that the dismissal was unfair for lack of evidence it may order the reinstatement of the employee with compensation for loss of earnings or it may award him substantial compensation as shown in the note to paragraph 4 of the ACAS Draft Code of Practice.

The third course is to call in the police to conduct an investigation and to be guided by them as to whether the evidence will support a prosecution. To suspend the employee with full pay pending a decision as to prosecution and to review the position regarding further payment of wages if prosecution is decided upon. In those circumstances the employer defers his decision on the disciplinary issue of whether or not to retain the accused in employment until the law has decided the issue of guilt or innocence of the offence alleged.

In reaching a decision about calling police a company which employs a well qualified head of a security department is in a favourable position because he will be able to advise them upon some of those issues and upon the strength and value of the evidence available to sustain a prosecution. He will also advise whether or not action by way of prosecution is likely to involve protracted court proceedings with resultant loss of company time for the attendance at court of witnesses.

Companies which do not employ such a qualified person may obtain the same kind of advice from one of the many people who are in business as security consultants and advisers and who will take a more dispassionate view of the problem than that likely to be given to it by police officers who may not easily recognise a company's point of view and best interests.

A desire to preserve the good name of the company by keeping that name out of court proceedings is fallacious – a company will seldom suffer harm to its business because it stands up to its public duty to bring wrong-doers to justice. There is a theory that every prosecution is a safeguard for future society by bringing an accused's weakness and wickedness to public notice and possibly in that way causing him to obey the law in future. In practice this is probably true in 70 per cent first offender cases.

The attitude of many company directors may be illustrated in the comment made by the managing director of one large business house to his security chief when he said, 'You may call police whenever you like so long as their action is not going to interfere with our business.' Nothing is likely to cause a security chief to sit back and ponder his position more than a remark such as that.

In the past several alternative lines of disciplinary action have been open to management for dealing with admitted or proven cases of theft by employees and these have ranged through dismissal; requirement to resign as an alternative to dismissal; suspension with loss of pay and service for a fixed period as a punishment for the offence; written or oral reprimand recorded on personal records; and suspension with full, reduced or no pay pending a decision upon disciplinary action. All of those alternatives are not open to management today.

Contemporary attitudes of the law and the courts towards offenders makes the old idea of sacking an employee caught stealing or even accused of stealing or otherwise offending against his employer out-moded and in many cases a mistaken action. Changes in that outlook have been brought about by the Contracts of Employment Act, The Industrial Relations Act, The Trade Union and Labour Relations Act 1974 and The Employment Protection Act 1975 which have progressively made it possible for employees to take action through industrial relations courts against their employers for unfair dismissal. Employers may also be sued in civil courts for compensation for defamation of character or other civil wrongs. Modern understanding of the law of civil rights coupled with free or assisted legal aid demands that employers must know where they stand in law before dismissing employees. If a dismissal is made for a breach of conditions of employment based upon an alleged criminal action which is not admitted by the accused and which has not been proven before a court then that dismissal is open to challenge as a lawful action and disciplinary action is generally inadvisable unless and until a prosecution through the courts has resulted in a conviction.

It must be noted that there is no bargain in the law about declining to prosecute because disciplinary action is being taken against an offender and at the same time there is no obligation upon anyone to take disciplinary action as an alternative to prosecution.

It cannot be too strongly emphasized that personnel records which relate to disciplinary matters must be carefully safeguarded in confidentiality because leakages can lead to further internal relations problems. Careless keyholding in this respect should in itself be a disciplinary matter calling for a recorded formal warning in the first instance.

Suspension of Employees from Work

Any consideration upon a declared company security policy should have regard for the question of suspending an employee pending the outcome of police investigation or of prolonged court proceedings. The alternatives are:

- i normal employment pending the result of an investigation then suspension if proceedings are to be taken and until a verdict is reached; or
- ii normal employment during investigation and between appearances on adjournment at court. It should be remembered that several months could elapse between the date of an offence and the date of final conviction or acquittal at a court of trial; or
- iii suspension without pay during the said periods or any of them; or
- iv suspension on full pay or reduced pay during the aforesaid periods or any of them.

It is suggested that it would not be right to apply one or other of those alternatives as a hard and fast rule and that a compromise should be found. Continuation in normal employment until the outcome of a trial could be an embarrassment to the company and to the security staff or to other employees and must rest primarily upon the circumstances which justified the calling of police in the first instance.

Suspension on full pay for any period allows too much liberty to the accused to supplement his normal earnings by casual employment. Any payment of wages so made cannot be reclaimed from him at any time.

Suspension without pay at any stage is tantamount to dismissal before the man has been proven guilty.

The middle of the road attitude will appear to be for a policy which permits normal employment or suspension on full pay during police enquiries but suspension at a reduced rate of pay from the outset of a prosecution and until conviction or acquittal. In the event of acquittal and reinstatement to employment the balance of pay between suspension rates and normal earnings must, by recent legislation, be paid in full.

The final word on this subject is that action by way of suspension would be inadvisable unless there is provision for it included in the terms of a contract of employment made between the company and its employees or in an agreed and declared policy as suggested by the ACAS Draft Code of Practice.

Retention or Reinstatement of Convicted Employee

This brings us to the question of whether or not an employee who admits to theft from the company or who is convicted in the courts of a crime against the company or any of its employees should be retained as an employee of that company or be reinstated after sentence is completed. Comment upon the company's attitude to this problem could properly be included in a declared security policy and aspects of the matter to be considered may include:

How much emphasis should be placed upon sentiment and personal animosity towards those who are known thieves?

How much emphasis should be given to the circumstances of the offence of which the person was convicted and its effects upon the operations of the company or of its employees?

Whether the company considers that criminal conviction should involve automatic dismissal. Companies which take into employment as new entrants persons with known previous convictions can hardly pursue automatic dismissal as a policy. Likewise companies who take newcomers into employment without verification of character references or (as in some cases) without requiring character references.

If this conviction has been for an offence involving less serious circumstances or has been this person's only wrongdoing, should he be reinstated in order to provide him with employment as an encouragement for him to keep upon an honest path in the future?

When the commission of the offence was in circumstances which had placed temptation in the offender's way by reason of a weakness in a control system or by a lack of proper supervision of a low paid worker, would it not be fair to attribute part of the blame for the offence to slackness on the part of the company and therefore to show to the offender some leniency?

How much emphasis should be placed upon the value of the man's skills and knowledge to the company?

If he is willing to go back to work amongst his fellows knowing that he will be the focus of attention, is there reason to suppose that he intends to 'live down' his offence and to behave himself in the future?

Is it fair to other employees to place amongst them a person convicted of crime and what are the views of the representatives of those employees upon this matter?

If a man has been punished by the due processes of law, is it in keeping with modern outlook to punish him a second time by depriving him of his livelihood with additional possible loss of pension and other rights?

In any debate about company security policy there will be those who consider that it is quite wrong for a court to have to take into consideration when passing sentence whether or not the accused will lose his job as the result of the offence before the court. If the court is told that the loss of job will take place the court is likely to impose a lower sentence than otherwise might be passed within the maximum authorised by law and as justified by the facts. It is felt by those who deprecate that situation that the courts should impose a proper penalty according to the facts and that that should be the only penalty without others exacting retribution by interfering with the accused's ordinary means of livelihood.

It is argued that in the days when stealing was a felony it was part of the common law that a convicted felon be deprived of his civil rights, responsibilities and privileges but over the past century such deprivation has fallen into disuse so that the Criminal Law Act 1967 finally abolished the division of offences into felonies and misdemeanours because that division no longer has practical application or value. Is it not therefore reasonable to assume that exacting retribution consequent upon conviction for theft should have ended with the passing of the 1967 Act?

Opposing voices in the debate will then bring forward the points of view that leniently dealing with one culprit encourages others to expect a 'soft line' from the company with regard to all offenders; that retention of a thief in employment makes accusations against other persons difficult to substantiate and that if a defaulter commits another crime someone in authority will be due for a rap over the knuckles for failing to supervise him properly.

All of what has been said in this Chapter with regard to theft may similarly be applied to other kinds of serious crime against the company or against its contractors and employees on company premises but there is yet to be considered the employer's attitude towards an employee who is convicted for crime committed outside the orbit of his employment.

Examples may be theft by shoplifting – concerned with others in serious assault or in a drunken brawl – found to be guilty of a crime involving indecency.

The modern point of view is that such a conviction does not necessarily make the person unemployable in his present job and if quoted as grounds for dismissal right be regarded as unfair. Exception to that attitude could apply if a man convicted of indecency towards women elsewhere works with a number of women or youths who object to his continued employment with them. In that case other work should be found for him as an alternative to dismissal. Dismissal on the grounds of theft elsewhere may be substantiated if the employee so convicted is engaged upon work which demands an extremely high level of integrity and dismissal of a fireman may be substantiated if he is convicted elsewhere of arson.

To summarize the main points of this Chapter:

Employers are urged by a legislative Code of Practice to formulate a policy and procedure for dealing with disciplinary matters and to make that policy and procedure known to their employees.

It is good management for a company to decide what disciplinary rules it will enforce and how it will enforce them and to make known its views on those matters to employees.

All alleged defaults should be promptly investigated and decisions taken communicated to those concerned in writing at the earliest practicable opportunity.

Losers of property or victims of other kinds of crime have an individual right to decide whether or not to prosecute offenders, they are under no obligation to do so, it is their public duty to do so, they are expected to support a case originated by the police or if the decision to prosecute is taken over by the police.

Procedure for a proper recording of, and investigation into, all known thefts and other crimes should exist in all businesses and losses may be reported to police by any loser or victim.

An employee who admits to theft of company property should be asked under caution to sign a written statement of confession and a decision may then be taken whether to prosecute, or to dismiss from employment or to take other disciplinary action which is in accord with the spirit of the Code of Practice.

An employee who denies an allegation that he has stolen company property should be the subject of police enquiries and no disciplinary action should be decided upon until the allegation has been proven in court.

Employees accused of crime may be suspended pending the outcome of a prosecution but it is good policy to write into conditions of employment that suspension with or without pay will be enforced in specified circumstances.

There are strong opposing views to be considered upon the issue of retention or reinstatement of employees who admit to or are convicted of crime.

Dismissal for a conviction for an offence committed outside the orbit of employment is inadvisable unless there are exceptional circumstances.

The preparation of a resolute policy drawn up in consultation with employee representatives and published for the information of all must result in good industrial relations and probably will have the deterrent effect of discouraging those who might be tempted to default.

It is far better to spend time dispassionately ironing out policy of this nature when there is not a case in hand waiting for a quick decision to be made than for a hurried decision to be made without a policy to back it up because the latter course is the first step towards a complaint of bad administration and towards friction between company and its employees.

CHAPTER THREE

Company Security Policy

Searching

Another matter about which a firm policy decision should be made is that of searching –

a) the clothing and hand luggage of personnel,
b) privately owned vehicles,
c) commercial vehicles,
d) employees lockers.

It is a fact that by far the majority of people are honest by nature but it is a fallibility of human nature that of that majority group there is a substantial section who will fall for a temptation or an unlooked for opportunity to steal. Most of this section can be deterred from committing unlawful acts by the fear of being caught or found out and, since the primary aim should always be to *prevent* crime, then any action taken towards catching or finding out those who are in the act of committing crime must be the strongest deterrent action available to society and thereby the greatest crime preventive.

Searching carried out as a regular day to day routine provides that deterrent.

Every company which allows people to walk (or drive) out of its premises with the property of the company taken unlawfully is encouraging the immoral standards of living which society as a whole abhores. The directors of such companies live with their heads in the clouds in the belief that either they have nothing worth stealing or that their workforce (and that includes the management) is so honest that they should not be required to suffer the indignity of search. Ask any truly honest man or woman whether he or she will agree to a search which will never place them under suspicion and they will forthwith consent. It is those who make up the substantial sector of not being honest all the time and who may succumb to temptation, plus those who are inveterate thieves who will object to search as a random, non-selective activity. Because a random search policy is designed and intended to deter those objectors from crime their minority objections should not be heeded and they should be required to accept a search policy as a condition of employment. What is the point of putting coiled barbed wire on the top of the perimeter fence, of weighing every lorry load, of auditing every accounting transaction if the loss of profits so saved is being frittered away each time employees walk out of the premises?

If the argument against a search policy is that there is nothing to steal then it must be assumed that those who make that decision are not alive to the kind of commodity which is habitually stolen wherever it happens to be. A few examples are:

Insulated electric cable There are people for whom this item in even the shortest of lengths is like nuts to a squirrel – they must gather it up and hide it away because at some time they will have enough copper stranding to make the collection a saleable item at a reasonably good price. The more 'spare' cable there is the faster 'the squirrels' will work and in some localities one has only to disconnect a circuit even temporarily for most of the line to disappear. Replacement costs money and that usually affects the profitability of the concern not only by the value of the cable but by the greater cost of re-installation.

Cloakroom and toilet fittings and fixtures These attractive chromed items just cannot be resisted by the do-it-yourself homemaker, male or female, and quite apart from shower heads, mirrors, paper holders and soap dispensers, they will take the chromed domes from fixing screws and the pull handles from flushing chains, not to mention the soap. All good costly items to replace at regular intervals in the interests of the 'honest ones'.

Working tools and their attachments In some concerns it might be profitable for the employers to provide each employee upon joining with two kits of tools – one for work and one for private use because it will not be long before that situation is achieved with many of the workers having a full kit at home regardless of the few tools used upon their work and this particularly applies to some non manual workers who are manual workers at home!

Timber of whatever shape, dimension, classification or condition – there are few people who can resist an opportunity to take home a piece of wood. Some of it is cut, shaped and drilled to size and fit for its intended use, whilst other is 'just in case it will come in handy'. Some of it is genuine scrap but quite a bit of it is created into scrap so that it will be discarded at the opportune time.

One might almost hear the voices of readers as they say 'But those are only petty things', so well they may be but allow people to get away with those items unchecked and disaster can strike the stocks of raw materials and finished goods which go out at the same time. There is quite a moral in the old story about the gatekeeper who was so busy checking the pass-outs for firewood each day that he overlooked the wheelbarrows and new sacks in which the wood was conveyed.

If the argument against a search policy is that the employees are all carefully screened and therefore are all honest it should be noted how frequently the lame excuse is put in all walks of life that there are black sheep in every flock. Take a cross section of any community and there will be dishonest and immoral persons amongst them. Four wrongdoers out of five hundred can wreak havoc with a business if unchecked and working in collaboration and that is only 0·8 per cent which is ten times lower than the national average of offenders.

There are companies at which a search policy exists but does not operate because the security staff find the task unpopular and abrasive with the workforce. That situation may be brought about

i by too much familiarity between members of the security department and the workers; or

ii by an inaccurate use of their authority by members of the security department; or

iii because top management has not given the subject sufficient emphasis in their education of the workforce and does not give a lead by example through line management to shop floor employees.

If all employees from top management downwards accept with good grace that they occasionally will be asked to go into a search room and if that activity is carried out with a ratio of stops between management, clerical and manual workers then the 'unsavouriness' of the security man's job will disappear and every search will be carried out with goodwill and as a normal part of the day to day running of the business.

All searching should be by random selection so that no one feels that he is personally under suspicion when asked to go into a search room and he/she will not be regarded by others as under suspicion even when he/she is.

Every search should be carried out with sufficient thoroughness to show that if something is being unlawfully removed from the premises there will be a serious possibility that it will be discovered.

Carrying out a search policy demands scrupulous fairness on the part of those involved and there must be no hint or suggestion of discrimination against individuals. The policy should apply to all levels and grades of employees alike and a routine should be followed which gives no cause for offence to those being searched or criticism of those conducting the searches.

In order that a routine may be thoroughly understood it is useful to employ a mnemonic as an aid to memory for all concerned. The mnemonic is the word 'SEARCHING' and runs as follows:

Selection Select several persons at random to be individually searched and include in a group any single person under suspicion.

Enter a searchroom Each selected person should enter a room with privacy.

Ask three questions

1 Ask each selected person whether he agrees to being then and there searched. Agreement on a previous occasion does not infer agreement now. Search without agreement on each occasion will amount to assault.

2 Ask each selected person whether he requires a witness to the search to be present. If he does, arrange this and also for a second security officer or supervisor to be present.

3 Ask each selected person whether he is in possession of any article which he should not be taking off the premises.

Refusals Remind those who refuse a search of the conditions of their employment and encourage them to submit to a search. If refusal is continued record conversation – tell the person that facts must be reported to management and release him to go on his way. A mere refusal to be searched does not create suspicion sufficient to justify detention, the calling of police or any other action beyond the report. Management action upon receipt of the report should be by way of warning for a first failure to comply with this condition of employment and by dismissal upon a subsequent occasion.

The only variation permitted upon that action is when the person has been selected for search because he was under suspicion by reason of something that was known about him prior to the request for a search. In the event of refusal by that person he may be detained for enquiries on the strength of the prior information and the police should be called.

Consenting parties Those who consent should be searched thoroughly enough to make it evident that property unlawfully possessed will be found. Handbags, hand luggage and all clothing worn or carried should be included in the search.

Helpful ones should be thanked for their co-operation.

Identify property Item of property about which there is suspicion should be quickly identified and the person in whose possession it is found should be told that he must await management directions.

Never disclose to a person who has been under suspicion prior to the request for search the identity of the person upon whose information action has been taken. This is a confidence which it is imperative to respect in the interest of that person's future relationship in the business. It is of value to make it known that confidence will be respected by the security staff.

Guard against the discarding inside or outside of a search room of any item unlawfully possessed by those about to be searched.

The searching of women may only be carried out by a woman and it is better to employ a woman on a part time retainer basis plus payment for visits rather than to employ anyone from the premises for this work. If the numbers do not justify temporarily employing a woman from outside then female searches should be undertaken by a female of supervisory rank.

The greater part of what has been said on this subject so far relates to the searching of clothing and personal hand luggage of individuals leaving business premises but there will also be situations and circumstances in which similar searching of persons entering, or moving about the premises will arise. These will not however fall into the category of random searches and will usually be requested upon an explanation being given of the grounds and reason for the search.

The searching of privately owned vehicles should be carried out on a much more regular basis and the privilege of bringing private vehicles into premises should be conditional that fairly regular searching will take place. Those who do not accept that condition will doubtless find accommodation elsewhere for their vehicles so that objections to the activity of searching private vehicles should be minimal.

Even more regularly should be the searching of commercial vehicles entering and leaving works premises and this applies very much to locations where lorries and vans back up to loading bays which are adjacent to the street. The equipment vehicles of contractors working on site should always be searched and, unless essential to the execution of the work, should be parked in the proper parking areas and not permitted to remain 'next to the job'.

There is the further matter of the company's rights to have a search made of employees clothing and tool lockers and it should be a matter of agreement between the company and the workers representatives as to the circumstances under which that action is taken.
Points to be included in an agreement on this matter might be –

1 That the allocation of lockers in no way debars the company's authorized representative from examining the contents of a locker at any time.
2 That a locker will only be examined with the knowledge of, or in the presence of, the person to whom the locker is allocated for use.
3 A locker will not be opened in the absence of the person to whom it is allotted unless (a) his absence contributed to an inability to inform him that it is intended to open the locker and (b) the manager of the department to which the holder belongs is present and the holder's representative if that is in the agreement.
4 A duplicate set of locker keys will be lodged with the head of the security department who will authorize use of the duplicate key as follows:

i To be loaned against signature to the holder of the original key for temporary use until the original key is available for use. The duplicate key must then be returned to the security file.

ii To be used by a security officer to open a locker under the conditions referred to under (2) or (3) above.

It will be a breach of discipline for any other use whatever to be made of a duplicate locker key held in the security department.

To summarize chapters 2 and 3 a company's security policy should include –

1 A written policy of intent, agreed with workers representatives, as to the company's attitude towards employees who steal company or other person's property; and towards persons other than employees who steal company property or other property on company premises.

2 A rider to the above policy of intent as to the conditions and circumstances under which internal disciplinary action may be taken by way of temporary suspension from employment of those accused of crime.

3 A declared policy that all conditions of employment shall include –

i an agreement by employees to co-operate with a random search policy for persons leaving company premises. Any breach of this condition being dealt with by way of a first warning recorded on the employee's records with more serious disciplinary action for any further breach of the condition;

ii that privately owned vehicles entering or leaving the premises are liable to be searched;

iii that employees driving commercial vehicles into or out of the premises will permit the vehicle to be searched on any occasion;

iv that subject to specified conditions the company reserves the right to examine the contents of any locker allotted for use by any employee.

It is suggested that the time involved in discussion and the preparation of a comprehensive security policy as summarized will be well rewarded

By the efficiency with which a security department will function and by the high level of morale which will exist in that department.

By the deterrent effect that such a policy will have upon the opportunists and those who might be tempted.

By the encouragement it will give to really honest people to report discrepancies and malpractices which come to their notice.

By the time that will be saved in debating policy and the action to be taken when incidents are reported or come to notice.

By the measure of satisfaction expressed by union representatives at the existence of a firm policy which they can understand and which can be understood by their members.

Any management which has not so far operated a policy of search must face up to the question of how to introduce searching to employees and their representatives in a manner which is acceptable to them. To those managements the following observations may be helpful.

1 Publicity should be given to known losses in order to show that there is a problem to be solved.
2 Publicity should be given to unexplained losses or differences between input and output with emphasis given to the fact that at least part of that loss or difference *could* be attributable to theft since all other avenues have or are being explored.
3 In a written memorandum or standing order to *all* employees the company should state:

   a) that it is entitled to protect company assets and is obliged to improve the profitability of the business for the benefit of shareholders and workforce alike;
   b) that the primary object is to prevent losses by theft rather than to catch offenders and that it is to that end that the company will introduce a policy of search by random selection of persons when leaving the company premises;
   c) that a random search policy is a preventive action which places no one under suspicion and in fact keeps all but those who offend clear of suspicion;
   d) that on and from a given date a random search policy will be operated and that new entrants to the company's employment will be required to submit to occasional search on a random selection basis as a condition of employment. A refusal by any of those persons to be searched *may* jeopardise their continued employment with the company because they will have breached their contract;
   e) that on and from the above date a random selection of employees will be asked to submit to a search and the co-operation of all in operating the policy is asked for. It should be clearly stated that no person will be searched without his/her expressed agreement upon every occasion upon which he/she is stopped but that refusal on any occasion will be recorded and reported to management as an act of non co-operation and that this will be recorded upon the individual's personal file together with his/her expressed reason for refusal;
   f) that the random search policy will be applied to all employees

impartially and that the employees of firms contracting for work on the premises will be subject to the same random selection procedure;

g) that searches will include the personal vehicles of employees parked within company premises no matter what their status.

4 Publication of the memorandum or standing order referred to in (3) above should be via introductory talks to all levels of employees or to their representatives and those talks should include a clear explanation as to how the policy will operate ie selecting persons by a numerical count, or in random groups, or by a random selection of individuals by the chief security officer and his staff, or by other named persons.

5 Preparatory to the introduction of search procedures the most careful training should be given to the security staff to ensure a standard procedure which cannot be challenged because it operates unfairly.

6 Whenever searching is carried out, whether it be of people or of vehicles, it is essential that a clear record be maintained in a search register showing time, date and details of person or vehicle searched and with reference to any report or action arising from the search. By nature many people resent a search procedure although they will co-operate by agreeing to be searched and it is advisable that the search register should indicate a resentful attitude so that management may be properly briefed should that resentment develop into some more serious objection on a later occasion.

CHAPTER FOUR

Company Security Policy

Personnel Vetting

It is good security policy for any company to scrutinize most carefully all applicants for employment *before* acceptance and the signing of contracts of employment.

This scrutiny should include considerations of past employment history as well as private interests and activities so as to avoid the recruitment of persons whose character and conduct make them undesirable employees or a bad security risk. Enquiries into character may not be made through official channels because of the restraint placed upon that information by the Official Secrets Acts.

At this point of time any discussion or consideration of the character of others must have regard for the Rehabilitation of Offenders Act 1975 which became law in England, Wales and Scotland on 1st July 1975. The Act does not apply in Northern Ireland.

The objects of the Act are set down in the preamble in the following terms: 'An Act to rehabilitate offenders who have not been reconvicted of any serious offence for periods of years, to penalise the unauthorized disclosure of their previous convictions, to amend the law of defamation, and for purposes connected therewith'. Hence it will be recognized that the Act is designed to assist the rehabilitation of persons who endeavour to rehabilitate themselves by 'going straight' for reasonable lengthy periods after completing a sentence or punishment imposed for a lapse of integrity, honesty or good conduct.

The Act describes a 'spent conviction' as a conviction to which no reference may be made after a period of years known as a rehabilitation period. Section 5 of the Act tabulates the rehabilitation periods which are related to the sentence imposed by the Court of trial and not to the maximum penalty for the offence for which the conviction was ordered. Rehabilitation periods are conditional upon no further convictions during the said periods. Part of Section 5 is as follows:

1 The sentences excluded from rehabilitation under the Act are –

a) a sentence of imprisonment for life;
b) a sentence of imprisonment or corrective training for a term exceeding thirty months;
c) a sentence of preventive detention; and

d) a sentence of detention during Her Majesty's pleasure or for life, or for a term exceeding thirty months, passed under Section 53 of the Children and Young Persons Act 1933 or under Section 57 of the Children and Young Persons (Scotland) Act 1937 (young offenders convicted of grave crimes);

and any other sentence is a sentence subject to rehabilitation under this Act.

2 For the purposes of this Act –

a) the rehabilitation period applicable to a sentence specified in the first column of Table A below is the period specified in the second column of that Table in relation to that sentence, or, where the sentence was imposed on a person who was under seventeen years of age at the date of his conviction, half that period; and

b) the rehabilitation period applicable to a sentence specified in the first column of Table B below is the period specified in the second column of that Table in relation to that sentence;

reckoned in either case from the date of the conviction in respect of which the sentence was imposed.

Table A

Rehabilitation periods subject to reduction by half for persons under 17

Sentence	Rehabilitation period
A sentence of imprisonment or corrective training for a term exceeding six months but not exceeding thirty months	Ten years
A sentence of cashiering, discharge with ignominy or dismissal with disgrace from Her Majesty's service	Ten years
A sentence of imprisionment for a term not exceeding six months	Seven years
A sentence of dismissal from Her Majesty's service	Seven years
Any sentence of detention in respect of a conviction in service disciplinary proceedings	Five years
A fine or any other sentence subject to rehabilitation under this Act, not being a sentence to which Table B below or any of subsections (3) to (8) below applies	Five years

Table B
Rehabilitation periods for certain sentences confined to young offenders

Sentence	Rehabilitation period
A sentence of Borstal training	Seven years
A sentence of detention for a term exceeding six months but not exceeding thirty months passed under Section 53 of the said Act of 1933 or under Section 57 of the said Act of 1937	Five years
A sentence of detention for a term not exceeding six months passed under either of those provisions	Three years
An order for detention in a detention centre made under Section 4 of the Criminal Justice Act 1961 or under Section 7 of the Criminal Justice (Scotland) Act 1963	Three years

3 The rehabilitation period applicable –

a) to an order discharging a person absolutely for an offence; and
b) to the discharge by a children's hearing under Section 43(2) of the Social Work (Scotland) Act 1968 of the referral of a child's case;

shall be six months from the date of conviction.

4 Where in respect of a conviction a person was conditionally discharged, bound over to keep the peace or be of good behaviour, or placed on probation, the rehabilitation period applicable to the sentence shall be one year from the date of conviction or a period beginning with that date and ending when the order for conditional discharge or probation order or (as the case may be) the recognizance or bond of caution to keep the peace or be of good behaviour ceases or ceased to have effect, whichever is the longer.

There are seven more subsections to Section 5 related to rehabilitation periods for court bindovers, orders for supervision, conditional discharges and similar matters.

Particular attention is called to the comparatively short periods of rehabilitation in cases where young persons were under 17 years of age at the time of conviction. Under the heading of 'Effect of Rehabilitation' the Act provides by Section 4

1 Subject to Sections 7 and 8 of the Act, a person who has become a

rehabilitated person for the purposes of this Act in respect of a conviction shall be treated for all purposes in law as a person who has not committed or been charged with or prosecuted for or convicted of or sentenced for the offence or offences which were the subject of that conviction; and, notwithstanding the provisions of any other enactment or rule of law to the contrary, but subject as aforesaid –

a) no evidence shall be admissible in any proceedings before a judicial authority exercising its jurisdiction or functions in Great Britain to prove that any such person has committed or been charged with or prosecuted for or convicted of or sentenced for any offence which was the subject of a spent conviction; and
b) a person shall not, in any such proceedings, be asked, and if asked, shall not be required to answer, any question relating to his past which cannot be answered without acknowledging or referring to a spent conviction or spent convictions or any circumstances ancillary thereto.

2 Subject to the provisions of any order made under subsection (4) below, where a question seeking information with respect of a person's previous convictions, offences, conduct or circumstances is put to him or to any other person otherwise than in proceedings before a judicial authority –

a) the question shall be treated as not relating to spent convictions or to any circumstances ancillary to spent convictions, and the answer thereto may be framed accordingly; and
b) the person questioned shall not be subjected to any liability or otherwise prejudiced in law by reason of any failure to acknowledge or disclose a spent conviction in his answer to the question.

3 Subject to the provisions of any order made under subsection (4) below –

a) any obligation imposed on any person by any rule of law or by provisions of any agreement or arrangement to disclose any matters to any other person shall not extend to requiring him to disclose a spent conviction or any circumstances ancillary to a spent conviction (whether the conviction is his own or another's); and
b) a conviction which has become spent or any circumstances ancillary thereto, or any failure to disclose a spent conviction or any such circumstances, shall not be a proper ground for dismissing or exluding a person from any office, profession, occupation or employment, or for prejudicing him in any way in any occupation or employment.

4 The Secretary of State may by order –

a) make such provision as seems to him appropriate for excluding or modifying the application of either or both of paragraphs (a) and (b) of subsection (2) above in relation to questions put in such circumstances as may be specified in the order;
b) provide for such exceptions from the provisions of subsection (3) above as seem to him appropriate, in such cases or classes of case, and in relation to convictions of such a description, as may be specified in the order.

5 For the purposes of this section and Section 7 of the Act any of the following are circumstances ancillary to a conviction, that is to say –

a) the offence or offences which were the subject of that conviction;
b) the conduct constituting that offence or those offences; and
c) any process or proceedings preliminary to that conviction, any sentence imposed in respect of that conviction, any proceedings (whether by way of appeal or otherwise) for reviewing that conviction or any such sentence, and anything done in pursuance of or undergone in compliance with any such sentence.

6 For the purposes of this section and Section 7 of the Act 'proceedings before a judicial authority' includes, in addition to proceedings before any of the ordinary courts of law, proceedings before any tribunal, body or person having power –

a) by virtue of any enactment, law, custom or practice;
b) under the rules governing any association, institution, profession, occupation or employment; or
c) under any provision of an agreement providing for arbitration with respect of questions arising thereunder;

to determine any question affecting the rights, privileges, obligations or liabilities of any person, or to receive evidence affecting the determination of any such question.

Under the heading of 'Limitations on rehabilitation under this Act' Section 7 provides in part

1 Nothing in Section 4(1) shall affect –

a) the right of Her Majesty to grant a free pardon or to quash a conviction or to commute a sentence;
b), (c) and (d) the enforcement or pursuance of matters relating to the enforcement of a conviction notwithstanding that it has become a spent conviction.

2 Nothing in Section 4(1) shall affect the determination of any issue, or prevent the admission or requirement of any evidence, relating to

a person's previous convictions or to circumstances ancillary thereto –

a) in any criminal proceedings before a court in Great Britain (including any appeal or reference in a criminal matter);
b) in any service disciplinary proceedings or in any proceedings on appleal from any service disciplinary proceedings;
c), (d) and (e) in any proceedings relating to adoption, care schooling, etc of children and young persons.

Under the heading of 'Unauthorized disclosure of spent convictions' the Act provides by Section 9 –

1 In this section –
'official record' means a record kept for the purposes of its functions by any court, police force, Government department, local or other public authority in Great Britain, or a record kept, in Great Britain or elsewhere, for the purposes of any of Her Majesty's forces, being in either case a record containing information about persons convicted of offences; and
'specified information' means information imputing that a named or otherwise identifiable rehabilitated living person has committed or been charged with or prosecuted for or convicted of or sentenced for any offence which is the subject of a spent conviction.

2 Subject to the provisions of any order made under subsection (5) below, any person who, in the course of his official duties, has or at any time has had custody of or access to any official record or the information contained therein, shall be guilty of an offence if, knowing or having reasonable cause to suspect that any specified information he has obtained in the course of those duties is specified information, he discloses it, otherwise than in the course of those duties, to another person.

3 In any proceedings for an offence under subsection (2) above it shall be a defence for the defendant (or, in Scotland, the accused person) to show that the disclosure was made –

a) to the rehabilitated person or to another person at the express request of the rehabilitated person; or
b) to a person whom he reasonably believed to be the rehabilitated person or to another person at the express request of a person whom he reasonably believed to be the rehabilitated person.

4 Any person who obtains any specified information from any official record by means of any fraud, dishonesty or bribe shall be guilty of an offence.

5 The Secretary of State may by order make such provision as appears to him to be appropriate for excepting the disclosure of specified information derived from an official record from the provisions of subsection (2) above in such cases or classes of case as may be specified in the order.

6 Any person guilty of an offence under subsection (2) above shall be liable on summary conviction to a fine not exceeding £200.

7 Any person guilty of an offence under subsection (4) above shall be liable on summary conviction to a fine not exceeding £400 or to imprisonment for a term not exceeding six months, or to both.

8 Proceedings for an offence under subsection (2) above shall not, in England and Wales, be instituted except by or on behalf of the Director of Public Prosecutions.

Those broadly are the main provisions of the Act but they are varied by The Rehabilitation of Offenders Act 1974 (Exceptions) Order 1975 which came into operation on 1st July 1975 and which is an Order made by the Secretary of State by virtue of powers conferred upon him by Sections 4(4) and 7(4) of the principal Act. This Order does not vary the provisions of the Act as they affect industry and commerce but for interest they may be summarized as follows:

1 In Section 4(1)(a) and (b) of the Act reference is made to *proceedings* at which evidence of spent convictions may not be given and at which a person shall not be asked, but if asked shall not be required to answer, questions about spent convictions. Schedule 3 of the Order lists excepted proceedings for the purposes of Section 4(1)(a) and (b). It is a fairly long list which would affect the interests of a person against whom action is being taken or who is appealing against action taken against him as a member of a leading profession or before a police disciplinary board or under the law relating to firearms and explosives registration and certification: the Insurance Companies Act 1974; the Prevention of Fraud Investment Act 1958; or the right to be appointed as a teacher in a school or establishment for further education.

2 The Order states that none of the provisions of Section 4(2) of the Act shall apply in relation to

a) any question asked by or on behalf of any person, in the course of the duties of his office or employment, in order to assess the suitability –

i of the person to whom the question relates for admission to any of the professions specified in Part I of Schedule 1 to this Order which lists the leading professions of medical practitioner, barrister, dentist, veterinary surgeon, chartered accountant, midwife, optician, chemist and registered teacher in Scotland.

ii of the person to whom the question relates for any office or employment specified in Part II of the said Schedule 1 which lists official appointments in the courts, the police, the prison service, traffic warden service, probation service, teaching for further education, the social services, youth club and cadet forces leaderships.

iii of the person to whom the question relates or of any other person to pursue any occupation specified in Part III of the said Schedule 1 or to pursue it subject to a particular condition or restriction which lists certain regulated occupations which have little relationship with normal industry and commerce.

In giving effect to the Exception Order as above, a person asking questions about character of a person specified in one of the Schedules to the Order or for a purpose specified in one of those Schedules, shall tell the person being questioned that by virtue of the Rehabilitation of Offenders Act 1974 Exception Order 1975 spent convictions are to be disclosed.

Suffice it now to say that those modern requirements of the law make it abundantly clear that caution must be exercised not only in asking questions of applicants for employment but also when making written or oral enquiries to verify references. The Act does not restrict or penalize questions which relate to previous convictions but it does authorise a reply which omits a reference to spent convictions and does demand that amongst other things employment shall not be denied to a person on the grounds of previous bad character which refers only to spent convictions. Thus a question on a form of application may ask 'Have you ever been convicted of a criminal offence, if so state the date and offence for which you were convicted'. If the applicant fails to reveal in his answer that he has a conviction for which the period of rehabilitation has not expired then that fact may later be grounds for declaring any signed contract of employment void but if he does not enter in his answer to the question on the application form, a conviction which is a spent conviction, any subsequent contract of employment based upon that form must remain a valid contract.

An excellent and detailed appraisal of the Act is contained in *A Guide to the Rehabilitation of Offenders Act 1974* by Brian Harris, published by Barry Rose (Publishers) Ltd, Chichester.

Since reference to official records is unlawful in this country, business-men must find other ways of seeking out information about the conduct and character of those to whom they propose to offer employment.

Enquiries in each case should be as searching and exhaustive as circumstances demand and will be far more detailed for some kinds of employment than for others but it should always be remembered that bad employees do not lose that propensity merely by a change of occupation or change of working location and it should be as much a task for security to identify those persons as it is for them to identify those with criminal tendencies and pyromaniacs. Close co-operation between personnel and security departments is essential in this work and there should exist a complete trust with executives of both departments relying implicitly upon the knowledge and judgement of each other without demanding too many detailed explanations. If an executive forms an opinion from his enquiries that an applicant would be a bad risk there should not be acrimonious discussion as to whether or not the applicant should be accepted.

For most forms of employment character references from past employers should be insisted upon and verification of those references should be the subject of correspondence.

For employment which demands integrity and trust such as handling money or goods of high value and for employment in key positions where loyalty and straightforwardness are essential, references from householders not related to the applicant should be obtained and verified by personal enquiry.

Whenever possible documentary proof should be required in support of information provided by the applicant about his education, armed service, special qualifications etc and such material should be copied for attachment to the successful applicant's personnel records.

Companies may subscribe to privately compiled registers of the membership of organizations known to have a disruptive attitude towards industry and commerce. Subscribers may refer names to the registers and seek information from them.

The pool of potential employees for most companies is limited to a radius of a few miles and this fact suggests that close liaison between personnel officers and the heads of security departments with their opposite numbers in neighbouring business concerns would be most wise in order to exchange information of value to each.

Local and national newspaper reports of the proceedings heard by Crown Courts, Magistrates Courts and other official panels of enquiry provide very valuable information from which it is possible to compile a fairly comprehensive record of convicted persons resident within the dormitory area of a firm's potential workforce. The record builds up year by year and can be expanded by periodic exchange of information between company representatives. The system has weaknesses in that people change addresses and woman change their name upon marriage but there is no reason why an application form should not ask for a previous address or ask a married woman the date of her marriage and maiden name. In the event of the local record showing a recent conviction in the name of an applicant who has not disclosed the conviction upon his application form it is quite in order for the interviewer to say to the applicant, 'On (date) and at (court) a man who gave the name (give name as recorded) was convicted of (name the offence on the record) was that man you?' The interviewer must assess for himself/herself the applicant's reaction and response.

The question put in that way is an enquiry and not an accusation and there should follow no discussion of the applicant's reply except that if the answer is in the affirmative, it may be pointed out that there is a place on the application form for the conviction to be entered. It should also be pointed out that disclosure of a conviction does not necessarily involve rejection of the application but that an attitude of mistrust at the outset is not the best foundation for a good employment relationship to follow. When interviewing referees the same line of enquiry may be followed by saying to a referee 'On date, and at Court, a person who gave the name of . . . was convicted of an offence . . . would that have been the person who is the subject of this interview?' It is a question, an enquiry, it is not defamatory or slanderous and there should not follow any discussion upon the referee's answer.

Time spent in maintaining a record and in verifying references and other information entered upon application forms for employment or the membership of clubs, societies, professional associations and the like is time very well spent in reducing future security and/or industrial relations problems and the professional ability of senior security people to assist personnel departments in this important matter should be fully appreciated and utilized.

It has been said in earlier chapters that we live in changing times and perhaps it is not fully appreciated by society in general how much change is taking place in its attitudes towards offenders. There were times when a convicted felon was debarred amongst other things from voting and from joining HM forces but today with the stigma of felony removed from the statute books we are asked to forget felonious crimes by the passage of a few short years and to open any door in the world of industry to such miscreants whether or not we wish so to do!

CHAPTER FIVE

Company Security Policy

Management Involvement

Two major factors of a company's security policy must be the extent to which it directs and expects managers at all levels to shoulder their security responsibility and the extent to which it supports the work of the security department if there is one. If there is no security department then it is all down to the managers and supervisors and if they abdicate their responsibility then there is no security at all.

Acceptance of a security department does not necessarily mean acceptance by everyone nor does it mean acceptance all of the time nor in all that it does. The more zealously or persistently security duties are carried out the greater will be the resentment towards the department as a whole or against the individuals who are applying the pressure. That resentment may be from the few who oppose the very idea of a security department because they resent all forms of authority or it may be more widely felt and expressed because security action appears to contain an element of injustice or prejudice or bias. If resentment is confined to the few anti authoritarians there is unlikely to be very much wrong with the running of the department but if resentment is from a wider section of the working community then there would appear to be a need to look for a reason because good security relies so much upon good relations with the people amongst whom it is employed.

There is no place for personal animosities or individual feuds and on the security side self discipline demands compliance with an unwritten code of conduct founded upon courtesy, tact with forbearance. Security officers accept with fortitude that they will never be popular and accustom themselves to pleasing some of the people some of the time but never all of the people nor all of the time.

One of the outstanding changes which has taken place in society over the past twenty years has been the attitude of people at work towards their foremen, overseers and managers and in their intolerance of authority in any guise. Today there are so many people 'stepping out of line' that whenever someone is checked or corrected the response readily forthcoming is 'Why pick on me?' as though that were an excuse for whatever malpractice is being challenged.

At some places of work there is so much open hostility towards supervisory grades that, apart from giving directives concerned solely with the progress of the job in hand, a great many managers and supervisors deliberately avoid any other intercourse with their work people. There is indeed a wide gulf of 'them and us' on both sides.

In that atmosphere security officers are expected to maintain good industrial relations and to pinpoint those matters which they must challenge and those bits of conduct which for the sake of good relations they will permit to continue. Frequently they receive no support or backing from management for the action they do take and may even be accused of inefficiency and neglect of duty if they allow to pass some matter which should have been challenged by a supervisor in any case. Yet there are other managers and supervisors who treat all security action with animosity or scorn.

Amongst departmental and middle management there will be found those who prefer to 'run a tight ship' and who resent what they regard to be interfering action whenever security asks questions about their department or area of responsibility or challenges any member of it; there are others whose resentment of security intervention is borne of incompetence which security action frequently highlights; there is yet a third group who blame security for every defalcation arising in the company and deny any personal responsibility in their efforts to cover up an inability to control their men.

Internal discipline is a management responsibility and in many companies the board of directors delegates that responsibility to managers of individual branches, establishments, or of departments in the larger complexes. In turn those managers expect their assistant managers, section foremen, and overseers to report upon offenders and all breaches of company rules including of course offences against the law. Such matters as lateness and absenteeism, of overstaying meal breaks, of idling in cloakrooms and locker rooms and of leaving work before time are matters which security officers leave entirely to departmental supervisors but gambling, drinking, smoking, unruly conduct and dangerous horseplay are matters for security intervention and report to management as also are breaches of the law such as assault, criminal damage, theft and fraud. In all of those matters and irrespective of whether police proceedings are being instituted at the request of the company or of private persons the departmental manager should cause internal disciplinary procedures to be put into motion.

It is fair to say that security officers know their position in these matters and go out of their way to show that they are not exerting supervision over the workforce and yet much of the action they are obliged to take originates from a failure on the part of managers and supervisors to perform their supervisory work properly.

Promotion or appointment to lower, middle or upper management carries with it responsibility for supervision over the actions of others but many accept promotion with no intention of doing anything which is not directly concerned with the production or performance of the job in hand. They have in the past elected not to criticize misdeeds of their colleagues and see no reason in their promotional role to become accusers – they therefore look the other way or give words of quiet advice such as 'don't you do that when I'm around'; they completely ignore one of the most important functions of management which is to protect the company's interests in all matters which affect the profitability of the concern; they also neglect to comply with company wishes that they should give the security department their wholehearted co-operation and support and believe that the existence of a security department relieves them of responsibility for employees conduct.

Some managers are so concerned that they do not rock the boat of industrial relations that they take no action to prevent 'pilfering' or the taking of unlawful 'perks', which are terms they prefer to use instead of theft and stealing nor do they take any other action to bring to justice those who commit such acts. Their misguided ideas of appeasing the workers allows them to tolerate all kinds of abuse of privileges and permits people to come and go at will and without question. Such weak management is aptly illustrated in the true story of the workshop foreman who rang the security office to say that one of this men had just driven his private car into the paint shop for a bit of private repair work and then added 'but please don't tell him it was me who told you – I've got to work with him'.

The peculiarity of the situation is that when a firm line is taken and an employee is brought to book he usually finds no friends amongst his colleagues many of whom deprecate his wrongdoing and express surprise that he should have been given the freedom he appears to have enjoyed.

Much of the inaction by managers and supervisors is due to lack of understanding of the measures which they could be taking to show authority, to reduce opportunity for theft or fraud, to remove temptation. to enforce paper controls and to exclude intruders and trespassers from places where they have no right or reason to be. Such measures include:

- i controlling the disposal of offcuts, trimmings and swarf much of which is highclass and valuable material;
- ii recording and properly disposing of packaging crates and cartons from incoming supplies, empty sacks, empty barrels and casks, pallet boards and similar residual material including goods returned as unsaleable because they are out of date.

Note – the absence of a properly organized recording and disposal system for goods referred to in (i) and (ii) provides workers with grounds for believing that they may dispose of such items to their own advantage as a 'perk' of their job and it is worth noting here the differences between 'perks' and 'theft'. A true perk is that for which consent to its appropriation by another has been given by the nominal owner; or is an item for which another may *reasonably* expect consent to be given for him to appropriate it.

Regular appropriation without challenge may lead to such a reasonable expectation.

Theft takes place when consent is not given nor is it likely to be given and when appropriation is therefore dishonest.

It does not take long for workpeople to assume that they have a right to take what they regard as perks when in fact every such transaction is theft because consent to the appropriation has never been given and would not be given if consent were to be requested. Middle managers frequently know what is going on in this respect and do not, because of a mistaken sense of fellowship, take steps to check it.

When the unauthorized taking of such items is challenged by a security guard, the guard is regarded as some sort of an ogre who is not in step with accepted practice and is referred to in such unflattering terms as 'officious'; 'overbearing'; 'jumped up'; 'big headed' and 'gestapo' when in fact all that he is doing is closing a loophole left by one or several others of management grade.

iii Middle managers should so organize the work of their departments that temptation and the opportunities for theft are removed by recording and placing under physical control overnight and at other times of absence, all half used cartons, sacks or crates of raw materials and all left-overs, over-runs and unpacked finished goods. Such good housekeeping simplifies the work of patrolling security staff because they can readily recognize a disturbance of a normally 'clean' area:

iv ensuring that paper control systems which are devised to audit the uses of materials within a department, or to audit movement between departments, are fully operated;

v advising employees against the folly of leaving tools unattended or of leaving personal possessions on workbenches when they could be safely out of sight in clothes lockers or bench drawers;

vi strictly following the routines of a passout system when approval is given for items to be taken home by employees;

vii checking that borrowed items are returned within a reasonable time and obtaining an explanation when they are not. Some checking in this direction is carried out by security based upon the passout forms they collect but such checking should not wholly be their responsibility;

viii strictly limiting access to his department to those who work there and challenging anyone else who enters or appears to be passing through, particularly if that individual does not so much as acknowledge the person in charge. Strictness in such matters will make others hesitant about entering and will encourage a patrolling security officer also to challenge persons seen to be entering or leaving. If comings and goings are a free for all, security staff will see no reason for stopping anyone entering or leaving rightly or wrongly.

These are but a few instances in which the work of management and security is inter-related and they illustrate circumstances which, if properly regulated by management, will reduce the occasions when employees are in conflict with security because they feel that security is trying to do a manager's job for him.

Other matters upon which managers could co-operate with or give support to security include:

ix managers ensuring that signatures which they receive from anyone are readable and if unreadable to insert on the document the name of the signatory in block lettering. At some future date security may want to identify that individual;

x managers checking that all copies of an original document are identical as to inserted detail so that copy no 1 in department X does not read 4, whilst copy no 2 handed to security reads 14;

xi that 'gift wrap' schemes are properly supervised and that 'spare' paper printed with 'vouchers' and 'offers' are disposed of under security conditions;

xii that check weighing is properly exercised to avoid frauds by those who do not completely discharge bulk containers or who arrive with underweight loads. The control of weighbridges is best carried out by employees of a security department, guards or civilian staff, who act as independent checkers and use the weighing procedure as part of the security screen at the gate;

xiii managers should be alert for persons causing damage to company premises and property and should intervene forthwith or call for a security officer to do so. Damage has been known to occur in order that its effect will be for half a shift to be sent home early whilst protracted repairs are carried out. Similar deliberate acts of damage are committed to increase ventilation in hot weather or to improve access to or egress from a restricted working area. The very fact that awkward questions are asked or that security is called in to make an investigation into criminal damage will show that such incidents are not allowed to pass unheeded and must be a deterrent upon a repetition of that kind of conduct.

xiv Co-operation between middle management and security is most essential in the enormous task of preventing the wasteful and unnecessary use of electrical energy, water, gas, fuel and heating of all descriptions and the unnecessary and wasteful use of company transport. Security officers have the prevention of waste as one of their primary aims but the part they play is a minor one when compared with the responsibility for preventing waste which rests upon middle management.

xv Some employees of private industry and commerce have spendthrift ideas because it is so easy to be generous with company supplies or when a customer is footing the bill. To curb those wasteful ways it has been found useful to impose paper controls so that every expendable item must be requisitioned from a controlled store and the simple fact that written request must be made for every item does appreciably reduce wastage and neglect in the use of stores items generally. It also eliminates the 'free for all' movement of stores items past security check points and so avoids one further point of friction between workers and the security department.

xvi On the fire front there is similar unwillingness on the part of some supervisory grades to enforce the rules and instructions essential for good fire precautions. Stacked goods are piled round fire fighting equipment or against fire exit doorways and in gangways and corridors and it not infrequently requires two or more requests from security before the overseer will cause them to be removed. If an accident occurs due to neglect or if there is delay in dealing with a fire situation those managers will offer the lame excuse that they trusted their workpeople to be more careful or more sensible and are not prepared to accept any blame for their own poor administration.

xvii The manager who prefers to keep security out and to deal with malpractices within his department himself should give due consideration to whether that is in the best interests of the company or whether he is building up further troubles for himself and the company by covering up now the actions of offending individuals.

A manager lowers his own standards by turning a blind eye to misconduct and would do far better to confide in the head of the security department who may possibly come up with a solution which will deal with a miscreant without causing the manager loss of face within his department and without exacerbating industrial relations for the security staff or for the company.

Managers and all supervisory grades do need to be given training in what to look for, who to look for, where to look and how to deal with what they find and some of that training can best be given by the head of a security department who will at the same time be improving his liaison with those he is teaching. Should a company prefer for such training to be given by someone other than their own security chief, the Industrial Police and Security Association can arrange for one or more speakers to be provided at nominal fees.

Much of what has been said so far has highlighted issues that are considered to be harmful to an harmonious place of work and there may be others which do not involve management at all but they are not so commonplace nor so easy to identify. On the other hand there are many matters dealt with by a security department which can and do create good relations between the department and the rest of the employees and these include the services given to the working community by those security officers who volunteer to qualify as first aiders. The amount of goodwill developed between the first aider and his patient is enormous and during the dressing of a cut finger there is a lot of bad feeling dispelled besides 'bad blood' and that does all parties a power of good. Even on those affluent sites which boast a medical centre with qualified nursing orderlies it is wise for security officers to be trained in first aid so that they can look after themselves when they are the sole occupants or can supplement and support the orderlies at a time of catastrophe.

Mention is made elsewhere of security office cloakrooms and in operating such a scheme the security staff is giving a useful service much appreciated by the law abiding majority of workers. Mention is also made elsewhere of a duplicate locker key service which is known to work smoothly and with approval. Informing car drivers of lights left on in car parks on dark winter mornings or of flat tyres or oil leaks are helpful actions which register with many people as do the passing on of personal messages, helping drivers to get cars started, minding children for sick employees visiting personnel departments, playing favourite records over the works broadcasting system and numerous other acts of service performed by a good natured security staff – and none of those things need detract from the more formal aspect of the security job.

Today many security officers feel the need to join a trade union and in some companies it is a requirement of management that they should do so. Many unions agree that security staff should not be required to take part in strike action so long as the duties which they do perform are related only to the protection of the premises and that includes the prevention of trespass, damage, theft and other crimes and the prevention of fire. Complications do arise when it is the security staff who withdraw their labour and to avert that difficult situation the head of the security department should use all of his powers of conciliation and appeasement. He is very likely to be aided in his efforts by the high standards of loyalty that exists amongst security officers and by their deep sense of duty. It is to be hoped that others will not seek to take an unfair advantage of those attributes.

CHAPTER SIX

Company Security Policy

The Protection of Premises

It has been said that the purpose of security is to enhance profitability by preventing losses by theft, fire, fraud, waste, damage and trespass and progress towards that end may be achieved by causing to be built into premises good fire and crime preventive features; by causing premises to be equipped with efficient protective apparatus; and by employing a trained and proficient security staff.

It is one of the fundamental principles of security that the aim is seldom if ever achieved by a single means, best results will probably come from a combination of all three suggested measures – average results will come from a combination of any two but only poor security and poor prevention will be achieved if only one of those means is adopted unsupported by either of the other two.

In assessing methods of prevention one is not confined to the mere physical stopping of access or of action but may include detecting devices upon the assumption that the risk of being caught is a deterrent upon wrongdoing and therefore a preventive action just as in fire precautions the installation of early warning detectors and fire fighting appliances is a prevention against the growth of fire. A factor to be borne in mind is that the major enemies of an intruder are tidyness, quietness and good light.

The most secure building is one which comprises four walls and a roof with no entrances or exits but in order to make that building usable doorways and/or windows must be inserted and these threaten the security of the whole. Physical prevention therefore starts with windows and ventilators being located in inaccessible positions but this defence is immediately weakened by someone wishing to drain the water off the roof attaching a rainwater pipe to the outer wall close by one of those windows so there arises the need for security window fastenings and/or anti-climb paint for the pipe; or by someone wishing to provide employees with the means of ventilating their work area giving to them the facility to open and close the ventilators and this requires security equipment to lock the ventilator when the workers are not there.

Since doorways are essential for entry and egress they must be designed and constructed in such a way that unauthorized access may be prevented at any time and this usually involves a back-up arrangement of manpower or electronic controls to admit only those who are authorized.

These simple illustrations convey the complexity and size of the problem and immediately endorse the contention that structural features, equipment and manpower are inter-related and essential to each other.

In the specialized matter of fire prevention construction plays a most important part by the inclusion into the design of buildings such features as fire break walls and doors, the compartmentation of wall and roof voids, fire and smoke proof corridors and staircases, and by the use of heat resisting construction materials.

Preventive apparatus such as fire detectors, fixed and hand portable fire fighting equipment, alarm systems, and automatically operated fire break shutters improve good construction but all such matters are enhanced by fire patrolling carried out conscientiously and consistently by members of a well organized security force.

Security protection can be expensive but in no field is it more appropriate to say that one only gets what one pays for – there is no cheap way to good security.

The amount of protection needed must be related to that which requires protecting and it is quite useless to expect one aged man of small physique to prevent any attack by fire, force or fraud upon a warehouse stacked with suede coats unless the design of the premises is adequate and is supplemented by efficient stock controls, sophisticated detector and alarm systems and good quality fire fighting equipment. On the other hand to require eight feet perimeter walls topped by barbed wire enclosing solid brick building with all windows alarmed and patrolled by a strong security force would be utter waste of money if the buildings are used for the manufacture of unattractive commodities. Those premises would be adequately protected by a small gate-keeping and patrolling security force aided by essential fire precautions.

Where premises demand a low profile of protection there may sometimes be an isolated building or office such as a roundsman's paying-in office at a dairy or bakery depot or the drawing and design office of a paper mill where the strictest security is required and where all the aids of locks, bolts and bars plus electronic trap protection to give warning of intruders are essential.

In those cases there must be back-up arrangements of on-site security staff or direct contact with a contract security watchroom so that police may be alerted promptly upon the raising of an alarm from the premises. Similarly there is no point in having expensive fire detector systems using radar scanner or smoke particle detector or even a sprinkler system linked only with exterior bells. They must either operate a manned on-site security office indicator or a direct line to a fire brigade and preferably both.

A fact which cannot be ignored is that some manufacturing processes involve the use of raw materials or machinery of as much, if not greater, value than the finished product and that these need protection against the arsonist and saboteur. Another fact which must not be overlooked is that there is a ready market for stolen swarf, old sacks, used bronze bearings, offcuts of electric cabling and lightning conductor strip, roofing lead, cutting and welding plant and even waste paper in bulk and the unauthorized theft of these items can have a crippling effect upon small businesses where profit margins are narrow. Moreover lightweight foamed plastics, some synthetic fibres, rubber and all packaging materials of comparatively low value have extremely high fire potential which could wreck a building and plant in the very minimum of time so that there is no longer a place of work and of profit at that location.

There must also be the general consideration of risk from external sources and greater perimeter protection may be demanded by insurers if the adjoining premises are vulnerable to crime and/or fire. Hence the keeper of a small furniture shop may be told by his insurers that his premiums will be raised unless he spends capital upon strong physical protection for his shop which is broken into with some regularity by thieves intent upon attacking either the jewellers to the east, the sub post office to the west or the manufacturing furrier to the rear. He, poor chap, is not losing stock but his claims for damage repairs offend the feelings of his insurer's underwriter.

The building-in of security features must of necessity relate to individual buildings and to the purposes for which each is to be used. A sample of such features may include:

1 Underground or rooftop employees' and visitors' car parks with direct access for vehicles from and to the public highway and with a pedestrian route for the drivers taking them past a security check point into the building.

2 Adequate space off the highway for visiting commercial vehicles to be stopped and checked for documentation or physical search by security staff.

3 An interviewing room close to the visitors entrance so that strangers are given the minimum of opportunity to inspect the premises.

4 Mezzanine level rooms from which observation can be had over any large working area or over an area used by the public such as in supermarkets, exhibition halls, museums and art galleries, a banking hall or other concourse, or car parking, loading and unloading areas for vehicles.

5 The provision of enclosed walkways for use by employees moving between busy working areas and canteens to minimize the risk of fire or theft within departments being traversed.

6 Arrangements for locker rooms to be adjacent to washrooms and

toilets and where personal possessions may be safely housed during working hours. Windows of washrooms and toilets may be opaque but those of locker rooms should be clear and adjoining a populated place.

7 Soundproofing of walls, doors and windows to provide working comfort for those working with their brains and not with their brawn. This feature aids security during non working hours when sound or movement detectors may be switched on within the soundproofed areas to identify trespassers and those who wander from other working areas maybe!

8 Cash wages offices located in positions which are regarded to be the best from a security point of view rather than in the most convenient positions from the point of view of the workforce or management but otherwise entirely vulnerable to attack.

9 The arrangement of lighting controls which provide for immediate floodlighting of an outside area or whole floor of a building by security and operable from more than one location.

The selection of the right security equipment or apparatus is one for individual circumstances and it would be quite wrong to attempt in this book to assess or comment upon the qualities and advantages or otherwise of the vast amount of such material available on the open market.

An earlier chapter spoke of the security industry being a lusty youngster and many inventive minds and commercial organizers have co-ordinated their efforts to board the bandwagon of demand in this field. Suffice it to say that it might be a very good plan for the prospective purchaser to do quite a fair amount of shopping around when he may find a quite reasonably priced piece of equipment capable of doing precisely the same job as an expensive and complicated installed system involving heavy maintenance and servicing costs. He may also find that his salaries or wages bills may be profitably reduced by the introduction of some modern mechanical aid which at the same time increases the security of his operations.

Company policy and the insurers attitude may strongly influence the type of fire and intruder detecting and alarm devices to be installed and regard will have to be taken of the following questions:

Is it intended to provide only an outer perimeter defence or is there to be a secondary line of detection by internal trap devices?

Are some of the goods of such high value that they must be inside an alarmed strongroom, safe or cabinet the approaches to which are also alarmed?

Is it intended that the first detected approach of an intruder shall cause loud and alarming bells or sirens to be sounded a) to scare away the intruder and b) to focus attention upon the premises from neighbours, the police or passers by?

Is it intended that the forces of law and order shall have an opportunity of catching the intruder(s) by responding to a silent signal raised by an intruder's movements at the premises? If so will there be an audible alarm given after a fixed time interval and will that interval be adequate to allow police time to reach the premises before its operation?

Is it intended to use an alarm signal, audible or silent, to alert a full time security force on the premises and will there be an automatic duplication of that signal to the public fire brigade and/or the police?

Is the risk to be covered of sufficient value to warrant the expense of private land line connections to fire brigade, police or security company's watchroom or in the interests of economy will reliance be placed upon a machine to dial out the service required over the ordinary telephone system?

Does the company believe in a policy of prosecution no matter who may be caught by the aid of the installation or does it wish to exercise discretion according to the subsequently discovered circumstances and identity of those accused or under suspicion?

Having made firm decisions upon those policy factors the purchaser should be content to allow the installing engineers to select the most appropriate devices and pieces of apparatus and controls.

Circumstances do exist in which physical protection plays a very small part and where prevention of loss by theft depends very largely upon the alertness of individuals as in the case of theft by shoplifting but even in this field electronic and mechanical aids are now available to both prevent and to detect such losses. The newest equipment being used in this respect is the computer controlled stock record which instantly records every stock change by the mere passing of a 'pen' across a panel upon which has been recorded the stock number of the item being sold. Under this system there can indeed be small chance of stockroom fiddlers, of sales staff and customer conspiracies or of undertilling and similar cash customer frauds. The business so equipped has one vital need and that is for the preservation of its computer hardwear and soft wear from harm by fire and/or criminal damage or to the technical reader from arson and sabotage.

No paper upon the subject of the protection of premises would be complete without reference to the package deal contract arrangements offered to industry and commerce by security service companies which vary in size from the one man and his dog outfit to organizations with international connections and employing thousands of people.

From these companies it is possible to cover a great many security risks but they do tend to concentrate upon the provision of manpower plus equipment and dogs for the majority of guarding and watching duties. They contract to cover specified hours and one of the primary benefits of employing them is that they remove from the hirer the worry and responsibility of finding personnel to fill vacancies and absences of a regular security staff.

The most serious consideration is whether it is good security policy for a company to admit to its premises and to entrust the safety and security of its business to persons who owe them no allegiance as employees and with the company exercising no control over the frequency with which the hired personnel may be changed.

Those considerations should reduce in importance as mutual trust and confidence between the two organizations develops and this will depend entirely upon the standard of efficiency, service and integrity displayed by the contract service company. The better organized that company is, the more carefully it recruits and trains its personnel, the more closely it exercises supervision over all of its operators the more acceptable and satisfactory will be the arrangement to the hirer.

Much argument and agitation has been aroused in recent years to persuade the Home Office to support a policy of licensing for security service companies in order that an approved standard of conduct and ethics by them may be enforceable. Whilst recognizing the desirability for such companies to be run on the most straightforward lines with honest reliable employees the Home Office has not so far seen the need for putting this into legislation and places trust in the efforts of the majority of service companies to do voluntarily what the advocates of licensing would require them to do.

It is for each potential user of their services to satisfy himself that the company he chooses to employ is of a satisfactory standard and it may fairly be said that the one man and his dog outfit may be just as reliable and possibly more so than some of the larger but less well organized or less trained companies.

This chapter has sought to draw attention to factors of policy which must influence the quantity and quality of protection to be applied to a business whether by way of physical or sophisticated preventive or detective systems. The decision as to how much protection and at what cost is one which can only be arrived at by a careful assessment of the known or estimated level of losses; by a professional survey of the premises taking into account also the class of business in premises adjoining; by due regard to the vulnerability of the business from fire risk or from the attractiveness to thieves of the raw materials, plant or finished product; and by a consideration of what would be the outcome for the business in the event of a major catastrophe from whatever cause and whether such causes are preventable by proper protection now.

Advice and guidance towards making the best decision about protection should be sought from police crime prevention officers, from insurance surveyors, from one's own employed security adviser or security manager, from several representatives of security equipment companies and from one's internal auditors and accountants. Only by giving all of that advice a detached and unbiased reckoning will the head of a business arrive at a decision to protect that business at a level which will be adequate to meet the known or anticipated risks without panic spending or farcical inadequacy. In short he will be performing the function which has developed into the quite new profession of Risk Management or Total Loss Control.

Above all top management should not underestimate the value of employing reliable and properly trained security officers who are much more flexible, adaptable and helpful to a business than all of the mechanical and electrical aids one may muster. Given a good security environment and some of those technical aids to improve his efficiency a security officer will bring an immense bonus to the business by his personality and practical participation in the overall protection of the premises.

CHAPTER SEVEN

Company Security Policy

Protection of Confidential Information

1 Information about recipes, formulae, stresses, blends and design may remain confidential for generations being that magic something which maintains the success of a business. Other will o' the wisp information may today be at the peak of importance and highly secretive, tomorrow useless and available to the world – today worth a ransom, tomorrow of scrap value only. Every industrial organization has certain information which it would not wish its competitors to acquire and, though this varies from industry to industry, it must, in the main be centred round finance, research, marketing, sources of supply or future planning and projects. There is also a considerable amount of information which although not directly of value to a competitor still is of interest to unauthorised persons – either inside or outside the organization from which it emanates. In this sphere are certain aspects of personnel information, union negotiations, information concerning accidents and suchlike personal matters.

2 The term industrial espionage will not be used again in this chapter but it is the art of wresting information from the archives or from those 'in the know' and of catching information at peak value for the personal gain of the catcher or of others. It is an art which is not of itself a punishable wrong – not a crime in this country – and there is but little in the way of punitive action that can be taken against the industrial spy. What action there is may be summarized as follows:

- i It is a breach of copyright to copy without consent of the author any artistic work which includes paintings, drawings, engravings, photographs, designs, and the original work of draughtsmen. An injunction may be sought in the civil courts to restrain anyone from using a copy taken without consent and a breach of the injunction may be punished as a contempt of court. The author of the work may seek damages to compensate him for financial loss suffered by reason of a use made of such a copy. The originators may claim copyright to advertising slogans, brand names, procedural instructions, reports of tests or trials and other originals and thereby endeavour to protect their work.
- ii Trade marks, patterns and designs may be registered and inventions prescriptions and ideas may be granted patent rights. Any breach of

the conditions of registration or of the rules of the Patent Office may be the grounds for an action for damages by the registered user or patent holder.

Note: In any action for damages under this or the preceding paragraph it will be necessary for the plaintiff to prove that he had taken reasonable steps to protect his interests and that he has suffered damage in spite of the protection he has given to that interest.

iii In order to acquire details of an idea or design or any confidential data it is usually necessary for a document bearing the information to be removed from its resting place and that removal dishonestly effected with the intention of permanently depriving the owner of his possession of the document is theft contrary to Section 1 Theft Act 1968. Furthermore, if, instead of removing the document as above, a copy of it or of its contents is taken upon another piece of paper owned by some person other than the taker, then there could be theft of that piece of paper.

iv If it can be proved that an employee has given away or sold secrets of his employer's business he may be sacked because he has repudiated his contract of employment which implies a bond of loyalty and trust between employer and employee.

v Perhaps the most frequent unauthorized use of confidential information is made by ex-employees who pass on to others, or use for their personal business, information gained during their previous employment. Some constraint may be placed upon this activity by requiring those engaged for key and confidential positions to enter into a bond not to disclose company information during employment with a concern or for specified periods after leaving in very much the same way as civil servants, police and other government employees are restrained for life by the Official Secrets Acts of 1911 and 1920.

3 Since the controls upon the gathering and passing on of information outlined above are so limited in scope and often difficult to sustain it is essential that security measures should be taken within a business to protect company secrets. Leakage of information may occur through inefficient or inadequate security measures but more frequently through carelessness or lack of attention to the basic principles of security. In Civil Service and Armed Forces circles the principle relating to the control of information is contained in the phrase 'the need to know' and it is a management responsibility to ensure that confidential information does not pass from those with the need to know to those without that need in the course of and for the purposes of their work. This control will frequently be found to be annoying in its application and restrictive in its results. It must, of course, be more liberally interpreted in industry than

it is in the Government service but, at whatever level it is enforced, it is fundamental to good security.

4 Information can be acquired by unauthorised persons by three means:

i visual – ie the sight of apparatus, materials etc which would give a clue to future developments, new methods etc;
ii the passing of such information by word of mouth – either through overheard conversation (telephones or otherwise) or deliberate but unauthorized disclosure to a third party who has no 'need to know'.
iii allowing documents – reports, memoranda, drawings, manuals etc – to fall into the hands of unauthorized persons, even if only temporarily.

Protection against (i) demands physical screening of apparatus, of the use of misleading packaging and the distribution of false information to negate interest in a project but such measures will only have a temporary effect and should be recognized for their short term value.

The breaches emanating from category (ii) can only be eliminated by a rigid observance of the 'need to know' principle supported by a strong measure of self discipline among those who are properly in possession of such information. Care must be taken not to discuss such information on the telephone – even the internal telephone – since such conversation may be overheard either at the exchange or elsewhere. Similarly, meal times etc must not be used to discuss this type of information since all those present may not have a need to know and, in any case, waitresses etc can clearly not have such a requirement.

It is the obtaining of information by the means outlined in (iii) – ie the unauthorised sighting of documents – which is the most usual source of leakage and the one which strict attention to security measures can do much to eliminate.

5 The requirements for the protection of documents containing valuable information fall into five headings:

i **Classification**

The decision as to whether a document requires protection and, if so, at what level, is fundamental to good security since if classification is wrongly made all the ensuing security measures will either be inadequate, excessive or wasteful.

Clearly the only person who can determine what the classification of a document should be is the person who originates the document, since it requires protection from the earliest stages. To enable this person to make a correct decision he must know what information it is necessary to protect. This implies that at a high level, the decision as to what information requires protection must be made as a matter of policy and this grading *must* be available to persons who might have reason to originate documents etc containing that information.

Industry should not require, for the protection of its own information, as wide or as varied a series of classification as is used for the protection of Government or Services information. It is not proposed to attempt a lengthy analysis of this aspect of the subject here but, as a preliminary definition it is suggested that only three grades would be required and that their usage might be –

a) 'Restricted' – information which should not be given to the Press, information services or employees below a certain level without prior authority from the originator; this definition is capable of infinite variation in time and depth by the addition of other phrases eg 'Restricted' (until 1.2.67) or 'Restricted' – (not below Departmental Head level until 1.3.67) etc etc.
b) 'Confidential' – information, the unauthorised disclosure of which would embarrass management, endanger an advantage over competitors or jeopardise future negotiations;
c) 'Secret' – information, the unauthorised disclosure of which would cause serious embarrassment to management or might endanger the future success of the company.

Having decided the broad definition of the security classification it would then be necessary to apply them to the individual aspects of the work and to distribute the resulting lists of classified aspects to all those who might be in a position to originate documents containing such information. Once this has been done all that remains is to apply the rules of security to the information – in whatever form it is produced. These lists should be kept under regular and frequent review so that they may be adequate to deal with current conditions.

ii Protection during Preparation

All secretaries, typists etc engaged in the preparation of documents classified as confidential or above, should be of mature age and known by experience to be trustworthy and responsible company servants. It is wise for a list to be prepared showing who may be utilized for such purposes. This list should be kept under constant review. In addition they must be trained to appreciate the importance of security information and to comply rigidly with the security rules for the protection of classified documents.

All shorthand notebooks, drafts etc used in the preparation of classified documents must be treated as themselves being classified and therefore afforded the same protection as the document itself. This also applies to carbon paper, particularly when new carbons have been used, to the ribbons of electric typewriters and to stencils, to tapes etc if mechanical means have been used for recording the classified information and to spoiled copies or work partly completed.

As soon as any document contains classified information it *must* be marked with the appropriate grading. This should be done by marking the word 'Restricted', 'Confidential' or 'Secret' as appropriate at the centre of the top and bottom of each sheet. If a considerable volume of classified material is anticipated it is useful to have paper with these classifications printed in. If paper of this sort is not available then the classification should be typed – in black for 'Restricted' and 'Confidential' – and in red for 'Secret'. It is not sufficient to mark the top right hand corner of each sheet with the classification – it must appear on the top and bottom centre of each sheet. The only exception to this rule is the case of securely bound books with numbered pages where the classification can appear on the front and back covers.

All 'Secret' documents should be numbered and a strict check kept that no more than the authorised numbers are produced. If any classified document is to be photocopied, care should be taken in selecting the person to do the work and occasion may arise when it is necessary to supervise the work in order to ensure that spoiled or excess copies are confidentially destroyed. It is good security to introduce a system for recording and attaching to a file the names of all persons who have cause to handle classified documents during the preparation stages. It is sometimes said that the marking of documents with their security classification in plain language creates a danger since it tends to draw the attention of the inquisitive person to that document and its contents. It is argued that it would be better to devise some system of coded marking by reference number, or letter or even by colour. By this means it is suggested that the informed will know that the document needs protection while the attention of the less well informed will not be attracted to it. This is fallacious reasoning since if the general rules appertaining to the custody and care of classified documents are rigidly observed those who have no need to know will not – and should not – have the opportunity to see the document. Coded marking can only tend to introduce a measure of carelessness into the approach of those handling them.

iii **Protection during Distribution**

Once a classified document is prepared it will almost certainly require to be distributed to a variety of recipients either in the establishment or elsewhere. The protection given to those documents and the means employed for their transmission must be governed by a set of rules rigidly enforced. The following schedule provides a suggested set of rules. These differ according to the destination of the document and its classification. When the destination of the document is within the establishment it has been shown as 'Local', anywhere beyond that area has been shown as 'Other'.

Classification	Destination	
	Local	Other
Restricted	One envelope marked 'Restricted' Transmission by internal mail or hand No receipt required	One envelope – ordinary post No marking on envelope No receipt required
Confidential	Two envelopes – outer bearing no classification: inner marked 'Confidential' and addressed to recipient in person Transmission by internal post or by hand No receipt required	Two envelopes – outer bearing no classification: inner marked 'Confidential' and addressed to recipient in person Transmission by ordinary post or by hand No receipt required
Secret	Two envelopes – outer bearing no classification: inner marked 'Secret', addressed to recipient in person and sealed with a wax or wafer seal Transmission by hand of authorized messenger Receipt required at all stages	Two envelopes – outer no classification: inner marked 'Secret', addressed to recipient in person and sealed with a wax or wafer seal Transmission by hand of authorized messenger or registered post Receipt required at all stages

The above pattern can be adjusted to afford greater protection if considered desirable but less than that standard would be inadequate. If receipts have not been returned after a reasonable period immediate steps should be taken by the sender to ascertain that the document has in fact reached its correct destination.

iv Custody of Classified Documents

It is of little use taking precautions to classify documents which contain information requiring protection and to safeguard them during distribution if, having reached their proper destination, they are left available to unauthorized persons. It is in this manner that most leakages occur.

Classified documents should never be left exposed on desks in an office during the absence of the occupant. When not in use restricted documents should be stored in a locked drawer – either of a desk or a filing cabinet. Filing cabinets containing classified documents should have no indication of this fact on the outside.

Confidential documents should be stored in locked filing cabinets or safes whilst secret documents should only be stored in safes or in filing cabinets which have been strengthened by the addition of a bar through the handles of the drawers fitting into a hasp at the bottom of the cabinet and secured by a padlock at the top.

Keys of the drawers of desks, filing cabinets or of safes should never be left in pin drawers or other unlocked receptacles. Ideally all confidential or secret documents should be protected by a combination lock rather than a key lock. If this is done discretion must be used in selecting the combination which should never be compiled of car index numbers, birthdays etc which may by known to or guessed by an intelligent intruder.

If a large number of keys are involved it is a good practice to have a small key safe – fitted with a combination lock – installed so that those keys are not removed from the premises and, in consequence, cannot fall into the hands of unauthorized persons through loan, theft etc. If keys are removed from the premises they should not be attached to a tab giving any indication as to the location of the lock to which they belong. If any tab is attached to them it should show an entirely non-commital return address eg a police station.

v Disposal of Classified Documents

All documents containing classified information above the restricted grade including notebooks, stencils, carbon papers, drafts as well as completed documents must *NOT* be disposed of through the normal waste-paper bins but should be shredded or placed in a secure receptacle (ie one with a lockable lid with a slit in it) for future incineration or pulping by a responsible person.

Shredding machines capable of reducing paper to strips of 4 mm are not expensive. Once shredded the materials can be treated as ordinary waste.

Care must be taken to ensure that tapes used for classified

information are treated in the same manner as any other classified document and that the tapes are cleared or destroyed completely as should the tapes of electric typewriters.

6 Once company policy has been clarified on the lines suggested or otherwise it will be the task of the security department to aid that policy by the following action:

i Security should exercise a strict control over the reception of visitors – to record their arrivals and departures and the names of the persons or department they visit – to arrange for visitors to be escorted to and from the person visited and to ensure that visitors do not convey into the premises cameras or other recording equipment without the specific authority of top management;

ii Security should challenge the internal movement of persons into or about parts of the premises where they have no authority or apparent reason to be;

iii Security should undertake the disposal by incineration or by shredding of papers bearing confidential information or to supervise such disposal by others;

iv Security should ensure by physical examination that safes, cabinets filing systems, wall charts and all similar places containing confidential matter are properly locked during times when staff are not present and this should include periods of temporary absence as well as times when the premises are closed for business.

v When carrying out random searches of persons or vehicles Security should be on the alert for documents, charts, lists, photographs, drawings, models, samples and any other article which may be in the course of removal from the premises for espionage purposes.

vi Security should exercise careful observation and supervision of cleaners who are on the premises out of normal hours of business. This especially applies to contract cleaning personnel to ensure that they do not examine papers or materials unduly and do not admit to the premises anyone who is not strictly a member of the cleaning team.

vii Heads of security departments should make periodic visits to unoccupied offices and to all offices and research and planning areas out of business hours to identify confidential papers exposed to view and to liaise with departmental managers for such material to be given confidential cover in the future. Visits should also be made to board rooms and similar meeting places to take possession of notes and other literature bearing confidential data discarded at the close of meetings. Other measures which could be taken at meeting places include (a) the display in the room of small notices asking delegates to take away scrap notes or jotting pads or to place them into a ballot box type container in the room for disposal by security department; (b) the provision of writing materials which do not hold impressions of written notes after the copy hearing the notes has been removed; (c) advice to delegates that if they temporarily withdraw from a meeting, or if a recess is called, documents bearing confidential matters should be placed inside folders or brief cases or otherwise removed from view because it is not uncommon for attendants to use recess periods to clear away coffee cups or to clean ashtrays and otherwise to have opportunity to see figures or names upon exposed notes etc.

7 Many people have published their theories about the protection of information and about the many sophisticated systems and measures available to scramble telecommunication or to make temporarily invisible the written word, graph or diagram but most such measures will be unnecessary if staff are carefully selected and then taught the rudiments of security for dealing with confidential matters and the true value of loyalty to one's employer.

8 **Bribery and Corrupt Practice**

Industrial spying lends itself very much to the need for there to be collaboration between employees of a concern and outsiders which collaboration is encouraged or rewarded by the passing on of bribes or gifts or free holidays or similar tokens of appreciation. There could even be bribery within a concern when one party possesses information required by another for his personal or private purposes.

Persons who give, receive, or attempt to give or receive a bribe are liable as offenders equally with those who receive or solicit such bribes or gifts. The law on bribery is contained in several enactments the principal of which affecting industry is the Prevention of Corruption Act 1906 which makes it an offence for any agent or person in employment to accept, or to attempt to obtain from any person for himself or for any other person, any gift or consideration, as an inducement or reward for doing or not doing any act in relation to his principal's affairs or business, or for showing or not showing favour or disfavour to any person in relation to his principal's affairs or business.

It is also an offence under the Act for any person to corruptly offer, give or agree to give any gift or consideration to any agent or person in employment under like circumstances or for any like reason.

9 Security officers in some establishments are vulnerable to approaches for reward to obtain or to disclose information or to refrain from taking an action which could disclose an offence or identify an offender. Action by an officer so approached may fall into one of two categories. If the approach relates to an arrangement for the future the officer should make

it clear that he does not wish to be implicated and should endeavour to close the transaction. Then without disclosing his intention of doing so the officer should at the earliest opportunity make notes of all conversation which contained the approach to him and should report the incident without delay to his superior or to senior management. If the approach relates to an incident now taking place or imminent the officer should point out to the person making the approach that he understands that he is being asked to act or refrain from action which is contrary to the interests of his employer and should ask whether that was intended. Whether the answer to that question is 'yes' or 'no' he should tell the person involved that the facts will be reported to management and he should then obtain a witness to a continuance of whatever action he was taking which gave rise to the approach. That action should be immediately followed by a report to his security supervisor or to senior management that an attempt to bribe him had been made. If the person approaching him actually proffers money he should take it, mark it and immediately inform the person proffering it that he will report the incident forthwith.

CHAPTER EIGHT

Company Security Policy

Emergency Contingency Planning

Later chapters will consider fire prevention legislation and practical fire precautionary measures but the attention of those who direct company security policy is now drawn to the essential and urgent need for plans to be devised and put into action for minimizing the risks to personnel, plant and property in the event of fire or other catastrophe.

Contingency planning is closely associated with the cost effectiveness of the business and it is therefore only right that a fire contingency plan should be drawn up by a panel representative of financial, technical and administrative interests under the chairmanship of the Works Director. The panel could usefully include the chief engineer and/or the chief maintenance engineer supported by one or two specialist engineers such as the electrical engineer and computer engineer; the safety officer, the security manager and (where appointed) the works chief fire officer, the chief accountant or company insurance officer, the person appointed as chief fire marshall, one or more employees' representatives from the general workforce and co-opted heads of departments for discussion as and when matters affecting their departments are under review. The tasks of this panel should be:

a) to ensure that not only are the demands of the law with regard to escape routes and the means for giving warning of fire complied with but also that recommended standards set by the Fire Offices Committee or by the various Home Office Committees on fire safety are met; to ensure that suitable and sufficient fire fighting equipment is provided and properly distributed through the premises and that it is maintained in serviceable condition;
b) to draw up and to enforce adherence to a set of rules for preventing fire applicable to all employees. Such rules to include reference to fire and smoke doors being allowed to swing free without wedges or other contrivances being used to prop them open; keeping gangways and doorways unobstructed; reporting the use of fire fighting equipment on every occasion of use; keeping the approaches to fire fighting equipment and to fire call points free from obstruction; directions about smoking as to times and places where permitted or

prohibited; good housekeeping and the fire preventive value of no litter and of speedy attention to spillages and overflows.

c) to prepare standing instructions to all personnel as to the names and telephone numbers of persons to be immediately informed in the event of fire or similar emergency and as to other essential telephone numbers and/or addresses of key people who may be needed to assist eg fire, police and ambulance services, emergency and non-emergency numbers; salvage and restoration specialists; removal and storage contractors; company architects and surveyors; outside contractors regularly employed by the company for engineering or joinery work; electricity, gas and water undertakings, emergency and non-emergency numbers; neighbouring companies willing to assist in an emergency with names of individuals to be contacted; names, addresses and telephone numbers of insurance loss adjusters or persons in head office who will advise such people.

d) to ensure that information and training is given to every employee as to action to be taken in the event of outbreak of fire. The sequence of that action should be –

   i to raise the alarm in the affected area;
   ii to take positive action to call a fire brigade or other assistance;
   iii to tackle the fire with hand portable appliances if physically capable of doing so;
   iv upon their arrival to leave the fire fighting to trained personnel and to go to an evacuation assembly area.

   The system for calling the fire brigade may usefully be published in a 'fire prevention handbook' a copy of which is supplied to all employees. The handbook should describe all fire prevention arrangements and rules pertaining to the premises and should clearly indicate each person's place for assembly for roll call. The handbook could usefully illustrate and describe the uses and methods of operation of the various types of fire fighting appliances and automatic fire detector systems on the premises.

   Such detailed information placed into the hands of every employee needs to be supplemented by Fire Instruction notices distributed through all parts of the premises endorsing and summarizing the action to be taken in the event of fire outbreak and as a constant reminder of the location of assembly points and of the routes thereto.

e) to appoint fire stewards, marshalls or wardens, to organize and to supervise evacuation of personnel and to ensure the safety of plant, equipment and the contents of cash tills or wages offices or of valuable goods during evacuation.

f) to arrange for practical fire drill exercises to be held at intervals in order to learn and to put right for the future, all of those practical difficulties and problems which arise when theory is put into practice eg how many women will want to go back into an evacuated building to collect a forgotten handbag or top coat, or, how everyone overlooks the contractor's painter working on a fourth floor renovation and who ignores the alarm bell because he does not know its significance.

g) to designate the priority for departments, buildings or areas of the premises (a) for fire fighting and (b) for salvage operations.

h) to train selected personnel for salvage operations, to list materials required to aid salvage operations and to have such materials stored in appropriately accessible places. Those salvage aids might include plastic or other waterproof sheeting for covering plant, mops, brooms, squeegees, sacks of absorbent substance for soaking up water or for forming water channels or barriers. They could also include hoists, jacks, crowbars, shovels and forklift trucks, rubber boots and oilskin clothing.

i) to train persons who volunteer to become members of a search party in the event of information that a bomb has been planted on the premises. Persons accepted for this task should be those regularly working on the site and with mobile jobs which take them into various departments or areas. Forklift drivers, maintenance staff, stock supply distributors, janitors, permanent fireman, and many chargehands are amongst this group of suitable searchers. Each should know his assigned area or 'floor' thoroughly so as readily to recognize an unlocked cubicle or cupboard which is normally locked, or the 'normal' appearance of a cloakroom or canteen so that an unusual object is conspicuous to them. Searchers should be trained to go over their area daily or even several times daily to know the distinction between morning and afternoon conditions and to become so familiar with objects in, or removed from, their normal resting places that if something foreign to its surroundings has been introduced it is easily recognized. Should searchers find such an object during training rounds or during an emergency for which they cannot account they should not touch it but should report the finding to the head of security or other person responsible for the search operation whose job it will be to clarify the doubt or to call the police. Quite reasonably a suspect object found during training is less likely to be a bomb than one found during an alert but the exercise of recognition, of not touching and of promptly reporting may someday prove to have been worthwhile.

j) to survey buildings for structural changes necessary to minimise the spread of fire, to provide better fire breaks; to improve water drainage in the event of sprinkler operation or the use of fire hoses, to improve fire venting so that fire goes through the roof instead of spreading laterally; to improve ventilation to dispel smoke and smoke odour.

k) to plan for alternative and emergency short or long term accommodation for any works operation displaced by fire or other emergency damage; to plan for transport, communications, services and transfer of materials in such an event.

l) to plan for good communications with the workforce in the event of a major catastrophe; for prompt financial arrangements for the payment of wages; for good liaison with press and television services; and for a customer advisory service where necessary.

As a general basis for their work the panel should identify their problem by assessing and listing major losses which could be caused by a catastrophic fire or similar emergency. The list might include:

a) loss of the whole or part of a building or of several buildings with all that that involves in terms of losses of stock, plant and equipment, the loss of employee welfare facilities, the loss of records and administrative and operational information.

b) loss of power supply, fuel supply, means of communication.

c) loss of only raw materials and/or finished goods, or of component parts; or of special plant which is vital to a production line; or of records, plans, orders, data or information essential to smooth and continuous operations of the business.

d) loss of manpower by essential diversion to salvage and clearing up operations or as a result of injuries sustained during the fire etc.

Additional to damage caused directly by fire or explosion it is important for the panel to have regard to the fact that at the scene of most large fires (and indeed some smaller fires) much damage is caused by water and smoke.

Water damage

Damage from water may occur not only in the area in which it is normally being used, but also by water percolating through floors, down staircases and through service ducting, which could result in –

a) basement areas being flooded during prolonged periods of fire-fighting, such a situation often being aggravated by excess discharge from sprinkler systems and the fracture of storage tanks and pipes (sometimes due to valves and stopcocks being difficult to locate and/or operate);

b) water accumulating and spreading horizontally over floor areas, due to drains and scuppers being covered or blocked by floor coverings, stores or waste material;

c) carelessly sited stores, equipment, records, plans, documents being damaged. Examples of careless siting on materials and stores would be where they are standing directly on a floor, or stacked too high and/or being too bulky to allow protective salvage covers to be placed over them.

The following precautionary measures should be considered, with a view to minimizing water damage in the event of fire:

d) floors should be waterproofed where possible, particularly where valuable equipment is installed on the next floor below;
e) in the design stage, floor levels should if possible be sloped so that water flows towards the outer wall(s), or alternatively to wherever drainage is installed;
f) adequate drains, manholes and scuppers should be installed and clearly marked – where drains are not installed, sumps with automatic pumps are a possible alternative, to collect and clear surplus water;
g) emergency plans should include the use of roller towels, blotting paper, carpets etc to check the spread of water;
h) stores and equipment should where feasible rest on pallets (ideally pallets should be of non-combustible material);
i) stacks of stores should be of reasonable proportions for easy accessibility, with a space of at least 3 feet (1 metre) between stacks, ceilings, walls, etc so that salvage waterproof covers can be fitted speedily in an emergency;
j) a space of at least 3 feet (1 metre) must be left clear below sprinkler heads;
k) racks and shelves (preferably of non-combustible material) should not extend to the ceiling and should not be of a design likely to retain water, nor should they butt directly on to walls;
l) valuable plant and equipment should, where possible, be sited away from areas where there are processes or goods which might require large quantities of water to control a fire, or where seepage from floors above is likely;
m) particularly valuable plant and equipment which is not sited under a waterproof ceiling should be protected by a waterproof covering (eg polythene sheeting) when not in use, or during an emergency;
n) important papers, records, plans etc should be kept in fire-resisting cabinets, safes or strongrooms as these are normally waterproof;
o) electrical switch gear should be mounted on battens and not directly onto a wall so as to avoid being affected by water running down walls;
p) electrical motors should be mounted on plinths rather than resting directly on the floor.

Smoke damage

a) smoke can damage goods, equipment and buildings by corrosion, contamination and impregnation with offensive odours;
b) smoke is made up of minute particles in both liquid and solid forms suspended in gases comprising many different substances and compounds; the composition of smoke can vary widely, depending upon the nature of the materials on fire;
c) the spread of smoke particles through a building which is on fire is accelerated by heated gases, these being more buoyant than air; thus all floors above the level of the fire are likely to be affected by smoke and may become completely smoke logged.

Where practicable the following measures will limit the extent of smoke damage in the event of a fire:

d) smoke doors installed in corridors, stair entrances and lobbies and closed as a fire and smoke barrier at all times; they should NOT be propped open with wedges or by other means for improving ventilation in warm weather;
e) staircases should be ventilated by means of skylights and windows;
f) automatic dampers should be installed at strategic points on conveyors and in air-conditioning ducts;
g) in large single storey buildings, roof ventilators should be installed to accelerate smoke dispersal;
h) in basements, smoke outlets with covers should be installed so that the covers can be removed or broken in emergency to allow smoke to escape;
i) materials which tend to produce a high density of harmful smoke (eg cork, certain plastics, vinyls, polystyrene and rubber) should be segregated from other materials.

Salvage and Restoration

Salvage and restoration operations should be implemented as soon as possible after a fire and the appointed loss adjuster consulted and his advice followed. The following general salvage and restoration measures may be undertaken as appropriate in particular circumstances:

a) clearance of debris after recovering usable items;
b) water may be pumped out of flooded basements, lift wells, machine pits etc;
c) drying out of floors and the remainder of premises should be completed as soon as possible – most forms of heating normally in use may be used for this purpose, supplemented as necessary by mobile space heaters (where obtainable from fire brigade or local

contractors) provided that such space heaters are carefully supervised and not left operating unattended, as they constitute a fire hazard in themselves;

d) machinery which has been exposed to damp or water should be dried, coated with oil if necessary and covered with plastic sheeting if exposed to the elements (or moved to alternative accommodation);
e) stocks of materials, stores, etc which have been affected by the fire should be removed as soon as possible and where applicable dried out;
f) damaged roofing must be covered;
g) damaged buildings must be boarded up and guarded;
h) arranging for a de-odourizing process to mitigate the effects of smoke damage, using ventilating fans and electrically driven sprays which produce a de-odourizing vapour – this process has been found to be successful in the case of building fabrics and finished products and to make offices and other working areas rehabitable.

Finally on this subject and of great importance is the action to be taken to restore used fire extinguishers and to have alram systems re-set as a matter of urgency. Also for the sprinkler system to be refurbished and for good lighting to be supplied.

Site Plans

The head of the security department should endeavour to acquire a copy of the plans of any site and/or building for which he is responsible. Such plans or drawings to show – cavity walls, wall thicknesses, roof voids, perimeter markings, drainage, sewers, water lines, gas and electricity supplies and linkage of those services with adjoining land or premises.

Such plans or drawings will be invaluable in arranging search areas or in deciding vulnerability of the various parts of the premises – it will also disclose possible hiding places for men or materials.

At some time it may be essential for someone to be able to produce these plans or drawings at short notice to the fire brigade or to a rescue team and who better to possess and produce them than the security chief.

Conclusion

Many of the recommendations set out in this chapter are commonsense actions naturally taken in a great many commercial and industrial houses where the planning caters for emergency situations to be dealt with outside of normal business hours in the same smooth and efficient way as they would be if the premises were fully manned. There are however far too many businesses in which fire contingency planning is given little if any thought because the profit hungry proprietors or overworked, harassed and financially pressed directors do not appreciate that such planning can i avoid panic and save lives;

ii ensure efficient use of manpower at the scene;

iii ensure an efficient and prompt attack upon a fire at a stage when it can be controlled;

iv allow those responsible for early action to go ahead without delays caused by unnecessary recourse to or interference by others;

v minimize damage and losses;

vi provide for a speedy and smooth return to normal working whatever the extent of the disruption.

CHAPTER NINE

Company Security Policy

Safety at Work and Security Involvement

Since the coming into effect on 1st April 1975 of the Health and Safety at Work Etc Act 1974 company policy in the matter of safety at work has been taken out of the voluntary and into the mandatory area of company–employee relationships. Those good companies who for years have been considerate of the health, welfare and safety of their employees now find themselves under legal obligation to do those things which hitherto had been a valuable part of their administration and they find it difficult to understand the reason for the change. Others who have been less thoughtful of their workforce in the past now find themselves confronted with obligations which they regard to be irksome and an imposition upon their freedom to run a business in their own way.

The Safety Officer

For different reasons both groups require to appoint a person who can devote time to an examination of the new law and can organize matters within the company, and liaise with parties outside of the company, so that the dictates of the law are complied with. That person is most likely to be designated The Safety Officer who in the larger concerns will be a full time safety manager but who in smaller companies may be another administrator such as the Security Manager, Chief Engineer or Commercial Manager.

For reasons of administrative economy, some companies embody the responsibilities for premises security, investigator, fire prevention and safety officer under one job title such as Site Services Manager or Risk Manager or Plant Protection Officer, and they sometimes tuck other responsibilities into the job description including Insurance Officer, Welfare and Catering Manager or Site Cleaning Supervisor.

Such multiplicity of roles causes some disquiet in professional circles as to which is the prior role and as to which special skill or experience is the best background for such an appointment. Each aspect of such a job demands experience in special fields and seldom will specialized training and experience in all fields be found in one man so where does his main interest lie?

If the company point of view gives emphasis to fire prevention then the Services Manager they appoint will probably be an ex-fireman who will know very little about crime and crime prevention. If safety takes precedence over security in the eyes of the boardroom they will appoint a full time specializing Safety Officer probably with engineering qualifications, and will ask him to supervise the site patrol and gate staff and fire precautions arrangements. This appointee's main interest will be centred upon the safe working of plant and he will require deputies to take charge of the crime and fire aspect of his job content. If the directors feel the need for an investigator they will appoint an ex-detective of senior rank and will ask him to take in the company's responsibilities for fire precautions, safety and welfare matters which roles he may not fill very well,

Industrial safety is indeed a specialist subject which has grown from voluntary action on the part of employers to reduce and to keep to a minimum the number of lost working days due to accidents which occur at work. Safety Officers examine the causes of accidents and takes steps to prevent accidents from happening due to those causes in the future. They examine accident hazards and potential hazards and implement measures to counter them. They encourage employees to use safety equipment and protective clothing and they arrange bonus schemes and similar inducements to workpeople to work safely and accident free. More than this they are concerned with forward planning to ensure that a new working area is safe to work in and that new plant or a new project is safe to work with.

The Law of Health, Welfare and Safety

Company obligations for the health, welfare and safety of employees formerly contained in the Factories Act 1961, the Offices, Shops and Railway Premises Act 1963, and other similar legislation relative to mines, explosives, petroleum, the construction industry etc have been superseded by The Health and Safety at Work Etc Act which extends those obligations so as to protect persons other than employees. The Act imposes other obligations not only upon employers but also upon manufacturers, importers, designers and suppliers to ensure that articles and substances are safe to work with when properly used.

The Act comprises 85 Sections and 10 Schedules many of the provisions of which have been implemented by Statutory Instrument no 1439(26c) under the title of The Health and Safety at Work Etc Act 1974 (Commencement no 1) Order 1974.

Security Officers who are also Safety Officers will need to study those provisions because they, no less than other employees, must comply with the obligations which the Act placed upon them whether from the point of view of managers or of employed persons. Additionally security personnel may find themselves involved in making arbitrary decisions

a) to support a claim from the workforce that the provisions of the Act are not being implemented by their employers, or

b) to support management as a whole, or any of its executive officers in obtaining compliance with a demand embodied in the Act.

For those reasons the following edited resume of some of the Act is presented for information but of course a full interpretation can only be obtained by a reading of the relevant Sections and cross-references in the Act itself.

Section 1 summarizes the purpose of the Act –

a) to secure health, safety and welfare of persons at work;
b) to protect persons other than employees against risks to health and safety arising from the running of a business;
c) to control the keeping and use of explosives and other dangerous substances;
d) to control the emission into the atmosphere of noxious or offensive substances.

Section 2 under the heading of 'General duties of employers to their employees' reads as follows:

1 It shall be the duty of every employer to ensure, so far as is reasonably practicable, the health, safety and welfare at work of all of his employees.

2 Without prejudice to the generality of an employer's duty under the preceding subsection, the matters to which that duty extends include in particular –

a) the provision and maintenance of plant and systems of work that are, so far as is reasonably practicable, safe and without risks to health;
b) arrangements for ensuring, so far as is reasonably practicable, safety and absence of risks to health in connection with the use, handling, storage and transport of articles and substances;
c) the provision of such information, instruction, training and supervision as is necessary to ensure, so far as is reasonably practicable, the health and safety at work of his employees;
d) so far as is reasonably practicable as regards any place of work under the employer's control, the maintenance of it in a condition that is safe and without risks to health and the provision and maintenance of means of access to and egress from it that are safe and without such risks;
e) the provision and maintenance of a working environment for his employees that is, so far as is reasonably practicable, safe, without risks to health, and adequate as regards facilities and arrangements for their welfare at work.

3 Except in such cases as may be prescribed, it shall be the duty of every employer to prepare and as often as may be appropriate revise a written statement of his general policy with respect to the health and safety at work of his employees and the organization and arrangements for the time being in force for carrying out that policy, and to bring the statement and any revision of it to the notice of all of his employees.

(Subsection (3) applies to any place of work where there are five or more employees)

4 Regulations made by the Secretary of State may provide for the appointment in prescribed cases by recognized trade unions (within the meaning of the regulations) of safety representatives from amongst the employees, and those representatives shall represent the employees in consultations with the employers under subsection (6) below and shall have such other functions as may be prescribed.

5 Regulations made by the Secretary of State may provide for the election in prescribed cases by employees of safety representatives from amongst the employees, and those representatives shall represent the employees in consultation with the employers under subsection (6) below and may have such other function as may be prescribed.

6 It shall be the duty of every employer to consult any such representatives with a view to the making and maintenance of arrangements which will enable him and his employees to co-operate effectively in promoting and developing measures to ensure the health and safety at work of the employees, and in checking the effectiveness of such measures.

7 In such cases as may be prescribed it shall be the duty of every employer, if requested to do so by the safety representatives mentioned in subsections (4) and (5) above, to establish, in accordance with regulations made by the Secretary of State, a safety committee having the function of keeping under review the measures taken to ensure the health and safety at work of his employees and such other functions as may be prescribed.

(N.B. Regulations referred to in subss. 4 and 5 above have not been made up to date of going to print with this book.)

Sections 3 and 4 make provision for implementing the purpose outlined in section 1(b) above.

Section 5 makes provision for implementing the purpose outlined in section 1(d) above.

Section 6 provides that

1 It shall be the duty of any persons who design, manufacture, import or supply any article for use at work

a) to ensure that such article is safe and without risk to health when properly used;
b) to carry out tests necessary to ensure (a) above;
c) to make available information about such article to ensure that it is safely and properly used.

2 It shall be the duty of persons who design or manufacture any article for use at work to carry out research to eliminate or minimize risks to health and safety in the use of the article.

3 It shall be the duty of persons who instal or erect any article for use at work to ensure that nothing about the way in which it is installed or erected makes it unsafe to use.

4 It shall be the duty of any person who designs or manufactures any substance for use at work to ensure that the substance is safe to use and that tests and research are carried out to that end and that information is available to ensure safe use of the substance.

Section 7 enacts: It shall be the duty of every employee while at work –

a) to take reasonable care for the health and safety of himself and of other persons who may be affected by his acts or omissions at work; and
b) as regards any duty or requirement imposed on his employer or any other person by or under any of the relevant statutory provisions, to co-operate with him so far as is necessary to enable that duty or requirement to be performed or complied with.

Section 8 enacts: No person shall intentionally or recklessly interfere with or misuse anything provided in the interests of health, safety or welfare in pursuance of any of the relevant statutory provisions.

Section 9 enacts: No employer shall levy or permit to be levied on any employee of his any charge in respect of anything done or provided in pursuance of any specific requirement of the relevant statutory provisions.

All of the foregoing Sections were given effect as from 1st April 1975 by Article 2(c) of Statutory Instrument no 1439(26c).

Health and Safety Regulations based upon the framework set down in Schedule 3 and authorized by Section 15 of the Act, came into effect as from 1st October 1974 by virtue of Article 2(a) of the same Statutory Instrument.

Section 18 authorizes the Secretary of State by Regulation to appoint local authorities as 'enforcing authorities' who by Section 19 may appoint Inspectors whose powers and duties are set down in Section 20. Those provisions have now taken effect with the result that the powers of Factory Inspectors formerly outlined in Section 146 Factories Act 1961 will in future extend over a much wider range of premises and with wider interests for the health, welfare and safety not only of people at work but of any other persons who enter or use company premises or who might be affected by the operations of a business whether on the premises or elsewhere and are now stated to be as follows:

Section 20

1 Subject to the provisions of Section 19 and this section, an inspector may, for the purpose of carrying into effect any of the relevant statutory provisions within the field of responsibility of the enforcing authority which appointed him, exercise the powers set out in subsection (2) below.

2 The powers of an inspector referred to in the preceding subsection are the following, namely –

a) at any reasonable time (or, in a situation which in his opinion is or may be dangerous, at any time) to enter any premises which he has reason to believe it is necessary for him to enter for the purpose mentioned in subsection (1) above;
b) to take with him a constable if he has reasonable cause to apprehend any serious obstruction in the execution of his duty;
c) without prejudice to the preceding paragraph, on entering any premises by virtue of paragraph (a) above to take with him –

   i any other person duly authorised by his (the inspector's) enforcing authority; and
   ii any equipment or materials required for any purpose for which the power of entry is being exercised;

d) to make such examination and investigation as may in any circumstances be necessary for the purpose mentioned in subsection (1) above;
e) as regards any premises which he has power to enter, to direct that those premises or any part of them, or anything therein, shall be left undisturbed (whether generally or in particular respects) for so long as is reasonably necessary for the purpose of any examination or investigation under paragraph (d) above;
f) to take such measurements and photographs and make such recordings as he considers necessary for the purpose of any examination or investigation under paragraph (d) above;

g) to take samples of any articles or substances found in any premises which he has power to enter, and of the atmosphere in or in the vicinity of any such premises;
h) in the case of any article or substance found in any premises which he has power to enter, being an article or substance which appears to him to have caused or to be likely to cause danger to health or safety, to cause it to be dismantled or subjected to any process or test (but not so as to damage or destroy it unless this is in the circumstances necessary for the purpose mentioned in subsection (1) above);
i) in the case of any such article or substance as is mentioned in the preceding paragraph, to take possession of it and detain it for so long as is necessary for all or any of the following purposes, namely –

   i to examine it and do to it anything which he has power to do under that paragraph;
   ii to ensure that it is not tampered with before his examination of it is completed;
   iii to ensure that it is available for use as evidence in any proceedings for an offence under any of the relevant statutory provisions or any proceedings relating to a notice under Section 21 or 22;

j) to require any person whom he has reasonable cause to believe to be able to give any information relevant to any examination or investigation under paragraph (d) above to answer (in the absence of persons other than a person nominated by him to be present and any persons whom the inspector may allow to be present) such questions as the inspector thinks fit to ask and to sign a declaration of the truth of his answers;
k) to require the production of, inspect, and take copies of or of any entry in –

   i any books or documents which by virtue of any of the relevant statutory provisions are required to be kept; and
   ii any other books or documents which it is necessary for him to use for the purposes of any examination or investigation under paragraph (d) above;

l) to require any person to afford him such facilities and assistance with respect to any matters or things within that person's control or in relation to which that person has responsibilities as are necessary to enable the inspector to exercise anv of the powers conferred on him by this section;

m) any other power which is necessary for the purpose mentioned in subsection (1) above.

An important new innovation into British law is contained in the powers given by Sections 21 and 22 of the Act to Inspectors appointed under the Act. Under those Sections an Inspector who has cause to say that a contravention under the Act is taking place may, without recourse to a court of law or other enforcing authority serve a notice upon the proprietor or person concerned requiring him to remedy the defect or to cease an activity until the defect has been remedied. The full text of the Sections is given for information –

Section 21 If an Inspector is of an opinion that a person –

a) is contravening one or more of the relevant statutory provisions; or
b) has contravened one or more of those provisions in circumstances that make it likely that the contravention will continue or be repeated,

he may serve on him a notice (in this Part referred to as 'an improvement notice') stating that he is of that opinion, specifying the provision or provisions as to which he is of that opinion, giving particulars of the reasons why he is of that opinion, and requiring that person to remedy the contravention or, as the case may be, the matters occasioning it within such period (ending not earlier than the period within which an appeal against the notice can be brought under Section 24) as may be specified in the notice.

Section 22

1 This section applies to any activities which are being or are about to be carried on by or under the control of any person, being activities to or in relation to which any of the relevant statutory provisions apply or will, if the activities are so carried on, apply.

2 If as regards any activities to which this section applies an inspector is of the opinion that, as carried on or about to be carried on by or under the control of the person in question, the activities involve or, as the case may be, will involve a risk of serious personal injury, the inspector may serve on that person a notice (in this Part referred to as 'a prohibition notice').

3 A prohibition notice shall –

a) state that the inspector is of the said opinion;
b) specify the matters which in his opinion give or, as the case may be, will give rise to the said risk;

c) where in his opinion any of those matters involves or, as the case may be, will involve a contravention of any of the relevant statutory provisions, state that he is of that opinion, specify the provision or provisions as to which he is of that opinion, and give particulars of the reasons why he is of that opinion; and

d) direct that the activities to which the notice relates shall not be carried on by or under the control of the person on whom the notice is served unless the matters specified in the notice in pursuance of paragraph (b) above and any associated contraventions of provisions so specified in pursuance of paragraph (c) above have been remedied.

4 A direction given in pursuance of subsection (3)(d) above shall take immediate effect if the inspector is of the opinion, and states it, that the risk of serious personal injury is or, as the case may be, will be imminent, and shall have effect at the end of a period specified in the notice in any other case.

Section 33 contains provisions as to offences under the Act, extracts of which are as follows:

1 It is an offence for a person –

a) to fail to discharge a duty to which he is subject by virtue of Sections 2 to 7;

b) to contravene Sections 8 or 9;

c) to contravene any health and safety regulations or agricultural health and safety regulations or any requirements or prohibition imposed under any such regulations (including any requirement or prohibition to which he is subject by virtue of the terms of or any condition or restriction attached to any licence, approval, exemption or other authority issued, given or granted under the regulations);

d) to contravene any requirement imposed by or under regulations under Section 14 or intentionally to obstruct any person in the exercise of his powers under that section;

e) to contravene any requirement imposed by an inspector under Section 20 or 25;

f) to prevent or attempt to prevent any other person from appearing before an inspector or from answering any question to which an inspector may by virtue of Section 20(2) require to answer;

g) to contravene any requirement or prohibition imposed by an improvement notice or a prohibition notice (including any such notice as modified on appeal);

h) intentionally to obstruct an inspector in the exercise or performance of his powers or duties;

i) to contravene any requirement imposed by a notice under Section 27(1);
j) to use or disclose any information in contravention of Section 27(4) or 28);
k) to make a statement which he knows to be false or recklessly to make a statement which is false where the statement is made –

i in purported compliance with a requirement to furnish any information imposed by or under any of the relevant statutory provisions; or
ii for the purpose of obtaining the issue of a document under any of the relevant statutory provisions to himself or another person;

l) intentionally to make a false entry in any register, book, notice or other document required by or under any of the relevant statutory provisions to be kept, served or given or, with intent to deceive, to make use of any such entry which he knows to be false;
m) with intent to deceive, to forge or use a document issued or authorized to be issued under any of the relevant statutory provisions or required for any purpose thereunder or to make or have in his possession a document so closely resembling any such document as to be calculated to deceive;
n) falsely to pretend to be an inspector.

2 A person guilty of an offence under paragraph (d), (f), (h) or (n) of subsection (1) above, or of an offence under paragraph (e) of that subsection consiting of contravening a requirement imposed by an inspector under Section 20, shall be liable on summary conviction to a fine not exceeding £400.

3 Except as enacted in subsection (2) a person convicted of any other offence shall be liable

a) on summary conviction to a fine not exceeding £400;
b) on conviction on indictment to imprisonment not exceeding two years and/or a fine.

Section 37 enacts:

1 Where an offence under any of the relevant statutory provisions committed by a body corporate is proved to have been committed with the consent or connivance of, or to have been attributable to any neglect on the part of, any director, manager, secretary or other similar officer of the body corporate or a person who was purporting to act in any such capacity, he as well as the body corporate shall be

guilty of that offence and shall be liable to be proceeded against and punished accordingly.

2 Where the affairs of a body corporate are managed by its members the preceding subsection shall apply in relation to the acts and defaults of a member in connection with his functions of management as if he were a director of the body corporate.

Section 38 enacts: Proceedings for an offence under any of the relevant statutory provisions shall not, in England and Wales, by instituted except by an inspector or by or with the consent of the Director of Public Prosecutions.

Chapter 13 of this book contains a reference to Section 78 and to Schedule 8 of the Act which relate to changes in the fire safety laws so as to phase these out of the Factories Act, the Offices, Shops and Railway Premises Act and similar legislation into amendments of the Fire Precautions Act 1971.

Security Involvement in Safety

Reverting the reader's thoughts to the subject of general safety at work and the involvement of security officers in company policy towards safety, health and welfare, it is true to say that quite apart from whether the head of the security department, whatever his job title, has special responsibilities, the duties of all security staff must naturally be concerned with the safety of employees and visitors to company premises. For example, attention to site lighting so that work may proceed safely in good light without wastage of power or fuel is a general responsibility of security; preventing access by intruders is a safety measure because strangers can quite unwittingly leave behind dangerous situations; good patrolling to detect gas leaks or other hazards are accident prevention activities of as much importance as picking up stray offcuts, stranded packing wire, and broken glass; controlling the speed of vehicles moving on site and enforcing other traffic control rules are security matters directly concerned with site safety just as much as the enforcing of company disciplines which prevent dangerous horseplay, misuse of forklift trucks, misuse of water hoses, compressed air lines and fire fighting equipment.

Security officers also have a definite contribution to make equally in the interests of the company and of employees by making impartial observations at the scene of an accident at work and by reporting without bias what they observe and assess to have been the cause of an accident. It is of immense value to a Factory Inspector or to a compensation board or to a court of law for a security officer to give evidence of a circumstance which he noticed at the scene of an accident and which has an important bearing upon where responsibility for an accident lies.

Quite naturally managers and supervisors called to the scene have a special interest on behalf of the company to observe the situation from a supervisory point of view whilst again, quite naturally, colleagues of an injured party will view the scene only with his welfare in mind. The prompt arrival of a security officer could precipitate any intended action by any party to adjust the scene and therefore what is ultimately reported and weighed in the balance is a true picture of the facts presented by an impartial security officer intent only upon assisting any later board of enquiry to come to an accurate assessment. Such intervention by a security officer may result in some pretty raw feelings of resentment that he should have 'put in his oar' and upset the balance of evidence in somebody's favour but it could well happen that the very next time he did that the scales would be weighted in the opposite direction – so who would be thanking him then?

Conclusion

Companies have always had their common law responsibilities as employers to provide a safe working environment and to meet the reasonable requests of their workforce with regard to health and welfare or face civil litigation for a settlement of disputes. It would now seem that law enforcement agencies have taken on the mantle of the civil law to compel the reluctant ones to fall into line with the great majority of good employers who can and do measure up to their obligations without penal legislation. It is however reasonable to expect that because of the new legislation those obligations will lead to a demand for specialists to safeguard company interests and it could be that the persons appointed as safety, health and welfare officers in future will not have time for other areas of administration such as security and fire prevention!

PART II

The Role of the Security Adviser

Introduction

When, during the 1950's, top management decided to adopt a policy of security to increase profitability it was natural that they should look for people with some expertise in the fields of crime and fire prevention and detection and with some organizing and administrative ability to become the leaders of their new security department. The people they employed for those posts were retired or retiring senior police and fire brigade officers who knew the law and had practical experience not only of fire and crime but of supervising men and were possessed of an ability to organize and to take charge of emergency situations. Indeed management found that they had obtained an unexpected bonus by taking into employment responsible people able to advise on many aspects of security to which they had given no previous thought.

Since those early days there have been many changes of personnel filling the top security posts and all have endeavoured to aspire to the high standards of proficiency and efficiency set by those 1950 pioneers of the security profession.

Chapters 2 to 9 have included references to a head of a security department or to a security manager or to a chief security officer but it should be clearly established that the Security Adviser, to give him a collective title, should now possess the training, study and experience to enable him to advise and give guidance to his management with regard to any of the matters discussed in those chapters.

He should also be capable of organizing against other threats to industry whilst at the same time running an important department of his own.

Part II of the book therefore takes a look at some of those threats to industry and at some aspects of the Security Adviser's role as head of a department.

CHAPTER TEN

The Threats to Industry and Commerce

The threats to which the Security Adviser should give his attention may be broadly listed under two headings:

1 Those which affect the majority of companies and classes of premises and which relate to –

   i the general security of buildings and their contents;
   ii the security of cash in bulk on premises or in transit;
   iii the security of goods in transit;
   iv fire precautions;
   v safety measures;
   vi traffic movements and parking;
   vii prevention of internal fraud and commercial fraud;
   viii protection of confidential information;
   ix the threat of bomb damage for ciminal, political or reactionary motives.

2 Those which are additional problems affecting individual companies, industries, or classes of premises and which relate to –

   i security problems of the building industry;
   ii security problems of the retail trade as to shoplifting, theft by shop staff, point of sale fraud and credit control;
   iii the storage and transit of explosives, corrosives, gases and highly flammable substances;
   iv the security of computer hardwear and softwear including the threat from fire, fraud or sabotage in computer suites;
   v the special problems of Vandalism in local authority controlled areas such as schools, parks, swimming baths and libraries;
   vi the special problems of hospitals and institutions, of hotels, docks, mines and defence establishments, airports and seaports;
   vii the special problems of the iron and steel industry and of the non-ferrous metals industry, the cosmetics and perfumery trade, the electrical goods industry and many many more such specialist groups.

This book endeavours to comment upon the broader issues of security action with regard to the problems in list 1 but no volume could possibly deal effectively with the wide ranging aspects of security depicted in list 2. where there is plenty of good rich material for authors with specialized knowledge to launch confidential or semi-confidential bulletins, and publications for circulation within those specialist fields. Readers with particular interests in any of them must forage a little for information to complete their researches.

In his primary role of fire and crime prevention the security adviser needs to be a visionary looking into the future to foresee the risks which may arise and to take effective action to prevent the advent of those risks. He should, by himself and by inviting his staff to confide their ideas to him, examine every threat to his employer's business even to the extent of assuming the romantic, fantastic, barely conceivable possibilities and to plan against those things happening. He should accept the philosophy that there has to be a first time for everything and that a fire could start in the most unlikely place, in the most unlikely material and at the most unlikely time or that a crime could be committed against all the odds of security defences, human failings and the risks of exposure and punishment.

It is not sufficient in these days of space travel, new methods of transport, new sciences and new materials to assume that new fire risks or new criminal methods will not evolve and the most successful security chiefs will be those who foresee and forestall any new attacks upon their security preventive screen.

In his book *The Electronic Criminal* published by McGraw Hill, Robert Farr graphically describes the modern approach to fraud and swindling crimes adopted by thieves using the modern sciences of computers, teletape, transmission of graphics, teleprinters and jet travel. But a few short years ago such a book could not have been written so who is to say what new chapters may be added by the passage of one short decade. It is surely only to be expected that space clothing and the ideas projected from the small screen in science fiction films will give criminally minded persons ideas for committing old crimes in new ways just as surely as the fellow who wedged himself in a chimney when he tried to emulate a burglary enacted on 'the box'.

In order to illustrate some of the foregoing observations this may be the appropriate place and time to review some of the subjects contained in list 1 which are not dealt with elsewhere in the book.

Security of cash in bulk on premises or in transit

The crime of robbery is defined and examined in chapter 16 but a fairly accurate and common understanding of the crime is that it involves theft by the use of force applied to the person whose property is being stolen or to persons who seek to protect that property. Mention of the word robbery causes one's thoughts to be drawn to the security of cash being conveyed between company premises and a bank or vice versa. Advice about changing the route and times of transit are well known as are other hints relating to the employment of men and not boys or females; for there to be escorts in addition to the carrier; for the use of an alarm bag or other strong receptacle but not for such bags to be attached to the carrier by straps or chains. This advice is well and good provided that other precautions are taken to conceal the exercise by a clamp down upon information outside and within the company as to how, when and by whom an exchange of cash between bank and premises is carried out. The fewer persons who know the details the better. 'Careless talk costs lives' read a wartime slogan and this was understood and respected by the populace at large. 'Careless talk costs money' might well be the slogan in the contemporary world of security.

Arranging with a local taxi service or car hire business to provide a vehicle at 10 o'clock every Thursday morning for a journey to and from the bank will be fraught with risks due to the changeability of drivers employed by the hire company but add to that some uncontrolled press publicity concerning a wages dispute with 500 employees and it will not give the thugs too much brain pain to estimate how much is conveyed in that vehicle every week.

Permitting an in-house journal to feature an article about the nimbleness of the wages department in preparing wage packets for two thousand employees in time for wage payments to begin promptly at 12 noon every Thursday is another way of advertising the place and time at which the estimated payroll may be available to those bold enough to take it. In-house journals find their way into the most unlikely places and may be found in Newquay, Newcastle or New York, in public waiting rooms, public houses or public baths or in trains, tea bars or toilets. Where else do criminals need to look for such valuable information. They can of course look in the Sunday newspapers to read of the gate money taken at every football ground in the country but turnstile money is likely to be bulky and in small change and therefore unattractive to them. They can go to Companies House to read the details of wage bills and weekly cash turnovers contained in annual reports required by the Companies Act 1967 and they can chat up the salesgirls and office typists at local dance halls to elicit similar confidential information to which those girls should not have been given access.

Security's job is to discover those titbits of information before they are given publicity and to gag them if it is at all practicable so that thieves do not know where those attractive sums are regularly available. Robbery is not necessarily an out of doors crime and can be committed on private premises and of course in cash offices and similar places used for making up or distributing wages and where the greatest risk period is just after delivery of the money by carrier and whilst it is still in a compact and easily transportable condition. Special security activity before arrival of the money should be centred upon a close scrutiny of vehicles arriving on site from daybreak on 'money day' to flush out bandits in box van bodies or in disguise as dustmen, draymen or draughtsmen. The building should be searched and patrolled and all other measures taken to ensure (i) that the carriers do not walk into an ambush on company premises; (ii) that bogus carriers or others do not have access to the room to which money has just been delivered; and (iii) that an emergency arising in some other part of the building does not result in the money and those in charge of it being placed at risk unduly. It is necessary for special instructions as a standing order to be given to the staff in charge of money in bulk as to their action upon the sounding of a fire alarm or similar emergency. It is also necessary for security to put an emergency plan into operation in such an event.

What has been said with regard to wages money also applies (a) to other concentrations of cash which may be delivered to social security offices, post offices and other payout centres; or (b) on premises where cash accumulates as at sports stadiums, supermarkets and showgrounds.

Today there are so many risks involved with transferring large sums of ready cash that the exercise should only be undertaken by the expert contractors who are equipped, trained and insured to provide a cash in transit service efficiently.

Expensive equipment including protective clothing, secure vehicles and good communications, backed up by good training and organized escort services form a formidable front against attack and enable the cash in transit companies to operate with the minimum of losses. Only if an industry, public department, local authority or commercial organization can provide for itself a comparable service should they handle bulk cash conveyance. Nothing less will be good enough to meet today's threat of attack and employers are unwise if they place their staff at risk in this respect. In the matter of forestalling the unexpected and unforeseen attack it should be remembered that the first payroll snatch by helicopter has yet to take place in these islands. Could that be your payroll?

The education of weekly paid workers into accepting bank drafts in lieu of cash for weekly pay makes steady but slow progress but whilst opportunities for wages snatch incidents thereby diminish the need for greater protection to be given to those monies still being conveyed through the streets is accentuated.

Robbery has always been an ugly crime with but small variance between the methods used by highwaymen and footpads causing concern for Henry Fielding two hundred years ago and those used by wage bandits and 'muggers' of the twentieth century. Better education, high living standards and the welfare society of our time have little influence upon the low moral code which is inbred into certain levels of mankind the world over and against whom the rest of society must protect itself.

Security of Goods in Transit

This subject is extensive and applies to an extremely wide range of industries and businesses. Goods which go missing whilst on the move may be the one parcel snatched by boys from the tailboard of a delivery van whilst the driver examines his delivery sheet, they may be two boxes of bullion exchanged for boxes of lead on a flight from Tokyo to London or they may be a lorry load consignment of valuable skins or spirits or meat stolen by the hijacking of the driver or by the fraudulent switch of keys and identities in a securely locked depot or parking lot. Those extremes may be few in number but in between there are thousands of losses from railways, the post office, public road services and privately operated transport systems. The loss of the small package containing a machine part for which there has been an eight months wait can be far more serious to the consignee than the loss of the bullion or lorry load will be to the insurers and the prevention of such losses must be a major security function.

The theft of goods in transit may be an act on impulse taken on the spur of the moment as and when opportunity presents itself to a thief or thieves or it may be a strategically planned operation effected after weeks of observation, time checking and with the aid of forged documents and persons being placed in pertinent positions.

The bigger jobs are seldom committed on impulse and in those cases the disposal of the goods has usually been arranged well in advance of the crime and is effected before the loss of the load or goods becomes known to the police or to the owners of the goods or their insurers. When it is necessary to gain time it is not unusual for drivers to be locked inside premises without a means of communication or to be driven around the countryside tied up in the back of a van before dumping where there will be a reasonable chance of discovery by travellers or local residents.

Protection for large or vulnerable loads must include a clampdown upon chatter and tit bits of information being passed around for thieves to pick up in public bars, cafés, in shops and on public transport. Good security starts with the action taken to screen those who are taken on as employees and this is specially important when recruiting drivers. A new driver should not be entrusted with a load of any value until his employment papers are known to be in order, his driving licence has been examined, he has been photographed and he has produced references which have been verified. Failing to take those simple precautions means that a firm is prepared to take a chance that this stranger will not drive away with a slice of the company's profitability on his very first trip. The lorry or van may fetch a good price even if the contents are of small value.

Systematic spot checking upon all vehicles in a fleet should be made during journeys to identify illicit or unauthorized cargo on board (perhaps exchanged for some company goods loaded in excess of the manifest for this journey), to check diversions from approved or direct routes and to check upon the driver's stopover points. Valuable and vulnerable loads should be entrusted only to established drivers who are paid an expenses allowance sufficient to enable them to avoid the cheaper cafés, roadside halts and lodging houses.

The security profession as a whole can help to prevent losses of goods in transit in the following ways –

1 By being alert to obtain the index mark of cars, vans and motor vehicles to which persons are seen to go when they have been hanging about or appear to have been waiting for the arrival or departure of a company vehicle – and by noting a description of such persons.

2 By checking the packing and distribution of loads inside large box van bodies. Extra pallets and plywood or hardboard panels used for packing should be examined for concealment of unauthorized goods on the load. Extra bags or sacks of produce carried to replace broken bags should be carefully checked for any which may be in excess of those which are authorized.

3 By a satisfactory examination of lorry seals to ensure that those affixed have not been tampered with in such a way that they appear to be correct when they are not. By conscientiously checking seal numbers with documentation for both outgoing and incoming loads.

4 By reporting to management the sales or distribution on site or through the vehicle fleet drivers of goods of any description at prices below those of normal retail outlets or manufacturers recommended selling prices. It will be noted that mention is made in chapter 23 of barter systems to exchange company goods for attractive items stolen elsewhere.

Over recent years extremely active and useful work to reduce the theft of goods in transit has been carried out by such organizations as The Tobacco Trades Advisory Committee and the Vehicles Security Committee of the Road Haulage Association Ltd. Both bodies report considerable reductions in the losses of bulk loads due to greater awareness of the risks which they have been able to engender in the minds of managements and those responsible for road transport. Representatives of both organizations have travelled freely to give talks containing advice and guidance to companies and to groups with an interest in reducing losses by theft of goods in transit and they have addressed many security training courses and security seminars. Literature containing sound advice has been widely distributed by the speakers and the quality of that advice may be assessed by the following extract from a leaflet published by the Vehicles Security Committee of the Road Haulage Association whose permission for its reproduction here has been given.

'The considerable experience and work which has gone into dealing with the problem of vehicle security has been boiled down into 12 points which every operator should bear in mind. These are set out below.

TWELVE CARDINAL POINTS

1 Make every effort to ensure you are employing honest staff. Take up references over (at least) the previous five years and be suspicious of unexplained gaps. When checking references by telephone be sure to look in the Post Office directory for the number yourself. One supplied by a dishonest applicant could connect you to his accomplice. Operators should draw up a staff enrolment form to suit the particular circumstances of the company.

2 Until you have seen his driving licence and, where appropriate, any other special licence, and have in your possession his P45 National Health card, and photograph, do not allow a newly-engaged driver to take out a vehicle.

3 It is helpful to have a trouble-free cash bonus which can be held wholly or in part as a fine on a driver who does not observe your security drill.

4 Fit an immobiliser and/or alarm in as inaccessible a position as possible. Choose a device for preference which does not make it necessary for the driver to carry out anything but the normal procedure in order to stop his vehicle. Inspect frequently.

5 Starter or ignition switches, security lock keys – remove numbers and keep the keys to each vehicle on a ring which is welded so that they cannot be separated.

6 If a vehicle's keys are lost, change switches and locks. It is much cheaper than losing a load.

7 Drivers of vehicles carrying valuable loads should be advised not to get out of their cabs if stopped. Even if a policeman requests them to do so they should offer to go to the nearest police station. Bolts on the inside of cab doors give added protection against hijackers.

8 Vehicles should not be left unattended for long periods, especially at night. At no time should keys be left in an unattended vehicle. A stationary vehicle, with its windscreen wipers or indicators operating, gives a clear signal to any watching criminal that it is his for the taking.

9 Discourage drivers from using the same cafés at the same time each day, particularly where their vehicles are not parked within sight.

10 Invite drivers to report to the police any suspicious circumstances which they might see on their travels – such as transfer of goods from one vehicle to another without apparent reason, or the registration number of any vehicle which is persistently following them.

11 When disposing of a vehicle, remove the name of your firm so that a thief cannot use it to secure a load by false pretences.

12 Remember that the police, especially crime prevention officers at your local station, and the RHA Security Committee are available for crime prevention advice.'

Amongst other points of advice the leaflet contains the following general hints:

- i The object of security devices fitted to vehicles is usually to buy time and to make the work of the thief more difficult.
- ii Electrically operated security devices should be those incorporating their own power supply.
- iii Mechanical devices are usually more difficult to overcome than electrical ones.
- iv A security device should not require the driver to take any additional action to switching off his engine, applying the brake and closing his door. For this reason a steering lock incorporating the ignition switch commends itself.

The Road Haulage Security Committee operates the Vehicle Observer Corps which was started in the London docks area in 1962 and which has been so successful in reducing theft of and from commercial vehicles that the scheme was once described by a Commissioner of Police for the Metropolis as 'a perfect example of crime prevention co-operation between industry and the Police'. The Corps now extends its activities over all major industrial areas of the United Kingdom and operates eleven groups in London and twenty-seven others in provincial towns and cities. Enrolment in the Vehicle Observer Corps may be made at any area office of the Road Haulage Association or the Freight Transport Association or by writing to Mr G J Burrows, Roadway House, 22 Upper Woburn Place, London WC1H 0ES.

The loss of small consignments and packages in transit through private or public transport services is not so easy to prevent and great reliance must be placed upon the standards of internal security operated by general cartage companies, by shipping companies, by British Rail and the Post Office parcels services, at cargo handling centres, at ports and airports, at railheads, and at major sorting offices and interchange depots. Prevention of loss of such consignments begins with proper and efficient packaging and labelling of goods so that pilferage through broken wrappings or re-direction by switching or changing the labelling are difficult to achieve.

The artifice of enclosing a package inside an additional outer wrapping labelled to the thief's own address is one not easy to thwart but can be countered by boldly marking on each package an identification mark or serial number of which checkers are required to make a record at every exchange point. Marking each item in a consignment such as '1 × 6', '2 × 6', '3 × 6', etc will cause checkers to look for the whole series of 6 items and if one is missing the approximate point of loss will at least be established for claims purposes. Using a service which requires goods to be accompanied by documentation requiring checking and signatures at each change of hands en route increases the chances of the goods getting through unscathed and warrants the extra costs incurred by such system. Thus a recorded delivery service, registration and insurance procedures, and company risk rates do safeguard goods in transit and should be used for that long awaited bearing which is holding up production or for that parcel of printed matter upon which a whole sales programme hinges as well as for the high value high risk commodities.

Packing and despatch departments in many companies may not be regarded as areas for security intrusion but delving into some of the long standing methods routinely carried out there may result in the injection of new security ideas which will cut the claims for non receipt of goods by consignees by half and another good job of improving profitability by good security will have been achieved.

The Threat of Bomb Damage for Criminal, Political or Reactionary Motives

'Bomb found by cleaner on an underground train in carriage sidings!'
'Bomb explosion at the Ideal Home Exhibition'
'Bomb found under a car outside the home of . . .'
'Bomb blast in a Surrey pub'

Such banner headlines in the daily press afford every reason for everyone to be his own watchdog and for everyone to be specially vigilant and aware of the actions of others no matter what the environment nor however friendly the 'atmosphere' may appear to be.

'If you see a package unattended DO NOT TOUCH IT'
'Keep your luggage with you'
'All persons entering these premises today are required to submit to a body search and all packages and hand luggage will be opened and the contents inspected'
'This entrance is closed until further notice – please use the main door'

These and similar notices confront us wherever we go and they are a clear reminder that at least some authorities and business managements are taking active steps to thwart the bombers or at least to negate the effects of their villainous and mischievous activities.

The problem would appear to fall under two main headings:

1 General awareness that a bomber may strike at any time without warning.

2 Action to be taken upon receipt of information that a bomb has been planted.

Security departments everywhere must be alive to risks of bomb threat and must organize accordingly. Special measures may be taken to reduce the opportunity for a bomber to gain access to business premises but determined people will find surreptitious means to get in and it is for *internal* security measures then to identify them.

All employees should be trained to accept as lawful only those actions of colleagues with whom they work or who they know personally and should question the presence of or any suspicious actions of any other person at their place of work. Personal identification badges for employees and for all approved visitors will be helpful so that a stranger in any part of the premises not wearing an ID will be challenged by someone and then security can be called in.

Everyone must be looking for the odd character(s) not necessarily out of keeping with his/her/their surroundings, who puts down a brief case, or paper carrier bag or rucksack or innocent looking parcel and then walks away from it. Several instances of observant people finding and reporting packages out of keeping with their surroundings have resulted in the de-fusing of explosive substances which could have maimed and injured many people. That kind of awareness should be part of our daily lives especially in city centres.

As to the second heading, threats made over a telephone or received in the post that a bomb has been or will be planted on your premises, may be genuine or hoax. We live in a world where more and more people are trying to impose their will upon others and who knows who is bluffing.

To industry the disruption of stopping work for the evacuation of employees is a major interference of the business which smooth handling can considerably alleviate.

Be fair to employees and get them out immediately but at the same time set into motion a 'think tank' of top people to decide how genuine is the threat and over what period is it likely to extend. The contingency planning panel suggested in Chapter 8 would be the ideal body for this task. Police may be alerted but they will not make management's decisions for them – they may give advice and guidance from their knowledge of other such messages affecting the locality and they may assist with public address equipment to facilitate communications and evacuation but the final decision of stay out or back to work must rest with the management of the threatened premises.

The more efficient and effective the normal security routines the less likely are the possibilities of a genuine bomb plant – the more regularly that fire drills and emergency plan exercises are carried out the less will be the disruptive effect to a business of a bomb threat call – if search training by regular search teams has been carried out the delay in the return to work will be minimal – awareness of today's problems should have ensured that emergency plan procedures are part of the routine of running a business and if your organization has not such a plan then someone is due for a rocket. No man should be so foolish as to believe that he is immune from harm or that he is the world's leading amateur bomb disposal expert. Anything suspicious or just not accountable for at the moment should be treated with respect and the real experts called in to take a look with the aid of their gadgetry. It is a good plan to have evacuation routes 'cleared' by the search party *before* evacuation takes place to avoid channelling large numbers of people past a danger spot! Do remember that flying glass can cause more injuries than the actual blast of a bomb, therefore assemblies of people should be moved away from plate glass windows unless they are within an area of general evacuation being ordered by the Police.

It is well known that the great majority of bomb threat calls are hoax calls originated no doubt by mentally deficient adolescents who in their youth trained themselves to make such calls by wantonly calling out fire brigades and by reporting non existent accidents to the ambulance and police services. The difficulty is to know which calls must be treated seriously and which may be ignored. The pattern of response would appear to be that all must be regarded seriously until or unless some fact emerges which is a clear indication that the call is hoax. The action ordered to be taken must involve the security force and quite frequently will rest upon the considered advice of an experienced head of a security department. This is where all the planning, preparation and training for handling emergencies should glide into operation. If there has been no planning or little preparation and practically no training the result will be 'panic stations' and chaos with the workforce being given yet another chance to taunt an inefficient management and a tepid security department.

A strict control at all entrances and exits is the best of anti-bomb activities and if that is supplemented by good premises patrolling the opportunity for a genuine bomb plant is greatly reduced and must be limited to a circumstance which has given to the premises or to the top executives the sort of publicity which makes them the target for an aggrieved party.

We have not yet reached a state of anarchy in these islands when bombs are planted inside *properly protected premises.* It rests upon the security industry as a whole to maintain that balance of control and the importance of the security office, and of the work performed there by the staff, must be given due recognition and must be sustained.

Each and every bomb threat call must be fully investigated to establish whether the call originated from within the premises and if there is the slightest possibility that that could have been so no effort must be spared to make it clear to the perpetrator that a repetition of that conduct is likely to result in identification and prosecution under Section 2 Criminal Damage Act 1971 which carries a maximum penalty of ten years imprisonment. It is surely pardonable here to repeat that the greatest deterrent to crime is the fear of being caught or found out and it is security's job to engender that fear whenever hoax bomb threat activity appears to be the work of someone inside the organization or premises.

One final word of advice – whenever evacuation is ordered someone must make sure that new arrivals do not innocently walk in – that those turned out stay out until the general order for re-occupation is given and that during the period of evacuation no-one is given the opportunity, nor takes the opportunity of looting the tills, or purloining personal possessions or of obtaining information to which they are not normally given access.

These things may sound simple but they can be real hazards in this kind of emergency and will need more manpower and more strategic use of manpower than most people will imagine to be necessary.

CHAPTER ELEVEN

The Organization and Administration of a Security Department

Moving away from his many roles of consultant adviser to management, leader of a planning and organization team for emergencies, liaison officer for safety at work, safety of transport and the safe keeping of cash and other property, we now turn our attention to the major responsibility for a security adviser to organize and administer his own department.

The primary objects of an efficient security department are:

The prevention of loss of company assets by crime.
The protection of life from fire, accident or other hazard.
The protection of property from fire, flood, damage and trespass.
The detection of offenders if crime or offences are committed.
The preservation of good order and the maintenance of the company's prestige at all times.

It must be emphasized that the priorities of the department must always be in that sequence and that prevention and protection are the first aims.

The effectiveness of a security department in carrying out those primary objects relies firstly upon the co-operation of the workforce and secondly upon the fullest support and backing of the company.

When employees can see that the company security policy is aimed at safeguarding their best interests by protecting them and their working environment they will accept and co-operate with a security department which does not exceed the bounds of its authority.

For that reason the security adviser should ensure that there is a written standing order which spells out the bounds of security authority, which is the company's directive to the security department, and by which the company must stand absolutely. A well-written standing order must show that it is published by the company and not by any one department or person. It must name the individual who will administer company security policy be he a director, security manager, chief fire officer or otherwise and it must show the chain of command and reporting procedure. By setting those standards right there will be no grounds upon which anyone may intervene or vary any of the routine instruction of which the remainder of the standing order is comprised.

A standing order for a security department should define the outlines of the work to be performed by the security staff without giving in detail who, when and where. For example:

- i the general hours of duty without details of the roster which will be published within the department only;
- ii the system of clock control without schedules or shift working arrangements;
- iii the method of using R/T communication without content of coded calls or 'hours of silence' etc;
- iv holiday and leave arrangements without rosters;
- v broad outlines of duties to be performed in the office and on patrol as a list of matters requiring attention rather than information as to frequency or method of giving that attention;
- vi specified action to be taken in emergency by the officers in the gatehouse and on patrol without naming names;
- vii directions about reporting and recording incidents and events and other operational matters.

Having ensured those working guidelines the security adviser should next turn his attention to the establishment of individual responsibilities.

Every well-organized job should start with a good job description which details the scope of the work expected of the job-holder. For the purpose of this exercise we must exclude jobs in the security industry which are concerned only with the manufacture, marketing and installation of crime and fire prevention equipment because those jobs are filled mainly by technicians and salesmen whereas we are concerned here with members of security departments and security service companies whose jobs involve them in the personal protection of premises from fire, fraud, theft, waste, damage and trespass.

Delegated Responsibility

Responsibility for security starts in the board room and that is where it ultimately remains. Delegation of the work may be to a branch manager, a departmental manager and/or to a security manager whose objectives should be –

1. To plan the organization, recruitment and training of personnel for a security department which has as its aim the prevention of loss of company assets.

2. To provide a security service which will protect life from fire, accident and other hazard; which will protect property and premises from fire, flood and trespass; and which will prevent crime, maintain good order and generally protect the company's prestige in the working environment.

3 To promote the company's fire prevention policy and safeguard the company's responsibility under the law for maintaining a safe working environment and adequate means of escape in emergency.

4 To direct the work of security personnel in the best interests of good industrial relations.

5 To make provision for the investigation of breaches of the law and of company rules and for offenders to be identified and reported.

6 To administer such other functions which are not incompatible with the security function as may be delegated to the security department as a permanency or from time to time.

The efficiency with which those objectives are met and executed will be shown by the performance standards of those who are employed in the security department, by the absence of complaints, by the level of morale in the department, by the preservation of good order and the absence of crime on company premises, by the promptness with which offenders are identified, and by the speed and efficiency with which fire, accident or other emergencies are dealt with upon inception.

The Head of a Security Department

The Security Manager, Chief Security Officer, Chief of Security Police or Security Adviser as we have chosen to call him should be very much concerned with company security policy acting in an advisory capacity to his Board upon all aspects of security within the business. His recommendations to the Board will include the creative work of constantly reviewing and recommending improvements in the security function and in audit control systems. He should be concerned with security problems associated with forward planning of premises and methods of operating and he should maintain close liaison with all management levels in the business and with many contacts outside of that business. Such matters as company cash in transit or on the premises, the disposal of confidential waste, communications, alarm systems and other security aids will be his special responsibility. He should be charged with the duty of investigating crime and other incidents involving serious breach of company rules or orders or any suspicious fire or emergency situation.

Security Supervisors

The head of a security department may need to delegate operating responsibilities to section heads, shift leaders or plant security officers whose objectives should identify with those of their manager but with less emphasis upon creative work and policy but with particular emphasis upon daily routines and operational responsibilities.

The work of these job holders will be supervisory at first line management level and will be concerned with such routine operating matters as the preparation of duty rosters and the disposition of security duties. They will particularly concern themselves with the arrival and departure of all personnel and vehicles, and will supervise the work of security officers performed in the security office and whilst patrolling. They may be responsible for the payment of wages, particularly to shift workers out of daily office hours and will be answerable to the security chief for the whole security function during their tours of duty. They will control communications from the security office in an emergency situation.

Security Guards

The rank and file security officer, guard, patrolman or whatever his title will usually perform shift work or changeable hours as a member of a team giving 24-hour cover to premises on every day of the year. He will generally be supervised by a shift leader or other security supervisor, or, if he works alone or in a small team, directly by his chief security officer. His duties will be varied but primarily concerned with gatekeeping, doorkeeping or similar control work in an office environment; with the patrolling of premises for fire, accident and crime prevention; with traffic control; and with the preservation of good order. He may be assisted by such security aids as electronic alarm systems, CCTV and photographic equipment, mobile radio telephony, special lighting, public address systems and fixed or mobile fire fighting equipment. His area of patrol may be bounded by or crossed by public roads, a river, canal or railway, each of which will give him added responsibilities. He will undoubtedly have places of special vulnerability to which he must give extra attention and his job will be greatly affected by the hours during which the premises are operating or closed for business. He will usually alternate between the office duty and patrolling duties at two or three-hourly intervals during each shift.

Those broad pictures of security work do not touch upon the volume of detail involved and limitation on space does not permit enlargement here. Such broad outlines should be the basis for each man's written job description which should specify that the chain of command and therefore of reporting must be through security management to a board level director or general manager and there should be no bypassing of that issue. So frequently security problems arise from or affect company policy and may involve middle management therefore they should only be discussed by the head of the security department with his director or top manager who holds ultimate responsibility for security in the business. Civilian clerical or operating staff who work in the security department must be answerable to the head of that department – there may be no divided loyalties.

Recruiting

Following upon changes which may be made in standing orders, new commitments undertaken by his department, promotions, or expansions of his security force the Security Adviser may well find it necessary to seek new staff.

When recruiting men and women for a security department serious attention must be given to the kind of man or woman to be recruited. These will be no ordinary individuals to be given a few hours of training and then left to carry on a job with light supervision – they will be special people with special interests and aptitudes whose training started when they were children and will continue for a considerable time after appointment.

In the first place they must be people who have accepted training in self discipline in their early lives, who have learned to respect authority and the rights and privileges of others, who have learned to control themselves and their way of living so that they are not forever falling foul of the law, getting into debt, behaving in a disorderly manner in public or at demonstrations, nor being over indulgent to the point of drunkenness or recklessness with gambling or the opposite sex. These will be reserved, sensible people of good character, of good bearing and of pleasant personality, able to approach problems with reasoning and understanding, forbearance and tact. They will be clean living people of clean appearance, helpful and considerate of youth or age and possessed of an innate honesty.

When employers advertise for security people they do not gloat upon the glory and advantages of the job but upon the kind of individual whose application is most likely to succeed and they use terms such as –

> 'must be fit, alert, smart of appearance and able to act upon own initiative, will be required to exercise tact and diplomacy in dealing with public.'

> '. . . responsible position calling for considerable experience in all aspects of security.'

> 'The man we seek will be directly responsible to a senior director . . . he will have either considerable civil or military police experience in an investigatory capacity or substantial experience in industrial security. The job is a demanding one requiring extensive mobility. . . "

> '. . . the successful applicant will be a man who can work on his own initiative and must be of sound integrity.'

> '. . . and we are looking for a man between 25 and 35 physically fit and of smart appearance, of good education and character and sober habits. He must be prepared for shift duties and occasionally extended hours of overtime.'

> 'Applicants must be physically fit and the successful applicant is likely to be in the age range 40–50 and will have had considerable previous experience of security work in the industrial field. Membership of the Institution of Industrial Security would be an advantage.'

How many other jobs demand such high standards of personality, character and experience as prerequisites for consideration and by how much do those standards limit the field for candidates? Employers are indeed looking for special people and should offer a salary which contains a reward for measuring up to those standards additional to any payment for work to be performed.

The shrewd interviewer of such applicants must first test the candidate for strength of character, for his or her ability to fit into the standards set and must test statements about schooling, interests, previous work and experience rigidly. Memberships of school cadet force, of team selections, of school expeditions, of private sport, social, athletic or drama interests must be questioned to distinguish the odd game now and then or the once only school playlet from the regular seasonal participation over a number of years, win or lose, and against strong opposition and other distractions. The interviewer must pursue the candidate's past history from childhood to yesterday and must be particularly suspicious of large unexplained gaps in the narrative. In the matter of previous work experience the interviewer needs to be much more interested in the reasons for leaving previous employment than in the type of work there performed and that interest should extend to a few telephoned or written enquiries through 'business contacts' whose personal opinions can be relied upon.

It goes without saying that any candidate to whom a security post is being offered must be prepared to take a medical examination before final acceptance. Age is important and those entering the security profession for the first time should be mature enough to know people and their ways and habits and therefore should not be under 25. They should be alert and active in mind and body and not so old that they are past their prime. Those with experience in the work may be accepted up to later ages and this very much depends upon how much previous experience and responsibility the job demands. Minimum height is sometimes specified for candidates but for this work personality and bearing are of greater importance because a man may be small but have a very impressive personality whilst a much bigger man may be just a jelly unable to command respect from anyone.

Security work is very much with and about people and those with personality are the leaders the world over so what we are really looking for in our candidate is an air of leadership. Maybe it is an ordinary patrolman that we are recruiting but every recruit should be looked at from the point of view of whether he could become a leader to take charge of a situation, to command respect, to show judgement and to make shrewd assessments. Only by recruiting the right people will the profession find its own future leadership and management.

Integrity, character, personality, fitness and physique all in one person are not easy to find but, having found him or her we then begin to ask about experience. Does this have to be ex-civil police, ex-military police or ex anything? If the post offered is in the lower echelon of the profession the answer is 'no' but if we want to fill one of the posts in the higher echelon the answer today most frequently is 'yes' because training within the profession in the past has been of such poor quality and so sparse that only a few of those recruited into the lower ranks during the security expansion of the past twenty years have gained sufficient technical and professional skill to be ready for command positions. This subject is further developed in Chapter 12 under the heading of 'Career Prospects in Industrial Security'.

Experience in security work is invaluable if it is available and can be found in applicants for posts in the lower ranks but one of the peculiarities of security work is that those engaged upon it are not usually 'birds of passage' but tend very much to settle to a steady job enjoying the benefits of regular employment. Therefore applicants with previous experience are not all that prolific and if one comes forward boasting of several years of varied security work with a number of employers he should be examined with particular scrutiny into his reasons for his habit of change.

Having recruited the right man or woman the security adviser must give serious thought to the training of that individual who must not be left to pick up only the good and bad habits of other security staff. Comments upon training may also be found in Chapter 12.

Other Administrative Matters

Other tasks with which the security adviser concerned with the running of a security department may have cause to give attention include:

- i to prepare and to frequently review a budget which is forward looking to the extent of one year, three years or even five years ahead;
- ii to participate in any management scheme for 'Management by Objectives' or similar target setting procedures in which each departmental head is expected to plan his work and the work of his department objectively – to fix target dates for achieving those objectives and to report progress and completion, or reasons for failing to complete, any of those objectives;

iii to make an annual report to his director upon each member of his department as to capability, suitability and interest in the work upon which each is engaged. The preparation of such reports may be preceded by a formal interview with each man and woman in the department or, such interviews may follow the preparation of reports when complimentary or adverse reports should be read to the interviewee. Annual reports may be associated with salary reviews – may be submitted in support of recommendations for salary changes and may make reference to bonus or other incentive awards;

iv to prepare for his board of directors at each year end an 'Annual Report' which summarises the work of his department for the calendar year. Comparisons with former years should be made – important events or incidents in which the department played a major role should be reviewed – changes in manpower, promotion, examination successes and general morale should be commented upon and the Report should conclude with a forward look to projects, changes, developments and pattern of work for the year ahead;

v to enquire the nature and character of new structural work or building alterations being planned and to project his ideas which will improve upon existing security features or will introduce new security measures at a lower cost than would apply if the work were to be carried out in isolation.

PART III

Training for Security

Introduction

Parts I and II have given a clear indication that aspirants for the top security jobs must be people of broad knowledge and experience and possessed of sound common sense. This applies equally to middle management in security being the senior security officers, shift leaders, branch security officers, inspectors or whatever title is used for those who are the number twos of the security world. They need to completely understand the law as it affects their daily tasks and to be proficient in the performance of their duties.

Security middle management must recognize that it is in the hot seat for setting a pattern of conduct to be followed by other middle management; for playing their part towards increasing profitability by following good security procedures; for controlling and directing the work of the rank and file of security personnel; for taking the lead into any investigation; and for setting up an incident control and reporting centre in the event of emergency.

Those tasks call for a great measure of training and Part III has been written for those persons in the hope that they will be inspired to broaden their outlook, to improve their technical expertise, and to merit further promotion.

There is also much in Part III to interest and to enlighten those security officers at guard and patrol levels who seek to improve themselves and to be considered for advancement in the progressive career which is available to them in the security industry. Only those prepared to make the effort should be considered for such advancement.

CHAPTER TWELVE

Security Training

The Need for and Purpose of Security Training

Training is described as undergoing instruction and practice to achieve a desired standard of skill and efficiency mentally and/or physically for a profession, art, sport or other purpose. In the security profession training is essential because of the wide-ranging but exacting nature of the job to be done. Gone are the days of the watchman, the gatekeeper and the doorkeeper. Modern security demands an intelligent application of a system which is a mixture of civil law, criminal law, company rules, employer's policy, union agreements, safety, health and welfare rules and arrangements and just plain common sense.

It is a system which is as impartial as it is resolute – as discretionary as it is forthright and requiring the patience of Job and the wisdom of Solomon. This is no ordinary job. This is a job which cannot be learned by watching somebody else doing it and no one person is capable of imparting to another all that needs to be known to be able to perform the job with efficiency and skill. This is a job for which fifty per cent of the efficiency and skill comes from the operator himself – his self discipline – his sense of fairness – his courage and his tact. To those qualities must be added skilled knowledge of the law, of rules and regulations, of paper commitments and of accepted practices and procedures. It is this latter group of skills which require training and no man should attempt to do the job or expect others satisfactorily to do the job, without formal training or study.

So many people do their job to a mediocre standard and make no effort to broaden their outlook or to adopt a more professional approach to the work. They are usually content immediately to hand over to someone else any matter of an unusual or contentious nature. Security officers who settle for such indifferent standards do not do justice to themselves, their employers or to the security profession. One may ask how such persons come to be doing the job at all and the answer is contained in the history and growth of the profession which has not yet developed to the stage where every man has been selected to meet the present day outlook upon, and demands of, commercial and industrial security. Despite several years of progress many of those employed in the security profession have received no formal training whatever.

Good training is best given by persons using their own years of expertise and we are indeed fortunate in the security profession in having such persons available and willing to impart their knowledge for the benefit of the profession as a whole. He who does not avail himself of that training, or who makes no efforts to obtain for himself that training, or who overlooks or neglects to arrange such training for security subordinates is indeed foolish because only by good training may a man be regarded as an efficient and skilled security officer.

Training for those already in the profession may vary according to the degree of previous training and/or experience of each individual and may follow one or more of the following patterns:

i *Internal training* within the company by whom the trainee is employed. By written standing orders. By memoranda and by written amendments or additions to standing orders. By verbal instruction from an experienced security leader. By practical work under the close guidance and supervision of an experienced colleague or security supervisor. By the study of security text books, manuals of security instruction, journals and periodicals devoted to security matters bought by the company for use by security personnel. By instruction classes arranged by the leader of a security department. By half-yearly or yearly briefing meetings for all security personnel under the chairmanship of the director/manager responsible for security within the concern. By participation in fire drills or fire fighting exercises.

ii *External training* by attendance at company expense and in company time at training courses arranged by professional organizations or technical colleges. By attendance at training exercises, exhibitions and demonstrations arranged by police or fire brigades. By attendance at conferences, seminars and symposia offered by police or fire brigade or by private organizations concerned with promoting business or management studies.

iii *Private study* by constant refreshing of knowledge via text books, journals and periodicals relative to security matters. By enrolling for a course of instruction by correspondence such as the courses offered by IPSA Correspondence Training College.

Important advances towards achieving satisfactory standards for the security profession have been made by the Industrial Police and Security Association who in 1969 obtained the incorporation of the Institution of Industrial Security offering membership in the grades of Fellow, Member and Graduate. Having taken up membership of IPSA as an essential first step towards recognition in the security industry and in order to become known to others the security officer may then look towards the Institution for status membership. The Institution offers qualifying examinations annually at Membership and Graduateship levels and successful candidates may seek Membership of the Institution by enlisting sponsorship and secondment from existing Fellows or Members of the Institution. Thus by personal contact and by examination standards of integrity and technical knowledge are carefully assessed before acceptance by the Institution as a Member of the Institution of Industrial Security (MIISec) or Graduate of the Institution of Industrial Security (Grad IISec).

Management in industry and commerce should recognize the significance of Institution status held by their security employees and by applicants for new security appointments.

Competition to gain Institution recognition is growing apace and it is the aim of this book not only to broaden the outlook and improve the technical knowledge of persons employed in the security industry so as to raise their general standard of professionalism but to assist students with their preparation for examination of the Institution and towards a MIISec or Grad IISec qualification to enhance their future careers.

Improving One's Security Knowledge

Mention has been made in earlier chapters of the broad scope of interests and subjects covered by the term 'security' and there comes a time when one may well ask oneself 'How broad is my security knowledge and my knowledge of security because these appear to be two different things?' Quite right, they are different things. *Security knowledge* is technical theory – knowledge of the law; of company rules and regulations; of theoretical routines to be followed in given circumstances; of the alternative lines of action open; of one's knowledge of equipment, its uses, advantages and disadvantages and of systems which are suitable in one circumstance but useless in another. *Knowledge of Security* is knowing how to apply and employ that technical theory; the understanding of security problems and the methods to be employed in solving them; the way to handle people and situations and an understanding of the practical approach which obtains results calmly and efficiently without embarrassment to the company or worsened industrial relations.

Security knowledge may be very deep indeed or extremely shallow according to the depth of study and sense of involvement of the individual. One person may be content to remember the basic definition of theft under Theft Act 1968 whilst another can quote dozens of stated cases by which the courts have interpreted the old larceny laws and will argue those decisions against the interpretations of the theft definition as given in Sections 2 to 6 of Theft Act.

One person may be content to know that a security officer who has made an arrest and prepared a statement for the police may be handed that statement (or a copy of it) to read and to refresh his memory before he is called into the witness box to give his evidence, whilst another person may not be content until he has studied the reported case of Regina v. Richardson in which the fundamental principle was argued as to whether or not the rules of evidence permitted a witness to read his statement at that time.

Some security officers like to become involved with the admissibility of confessions in evidence whereas for the majority it is enough that they get down into their notebooks what is said by an accused person as nearly as possible in the language he uses and to leave it to the lawyers to wrangle as to whether or not those notes may be referred to in evidence.

There are those who argue that without a deep and thorough study of the theory a security officer cannot get his knowledge of security right and that sound practical work is only possible against a sound theoretical background. There are others who assert that so long as the top man knows what he is doing there is no need for anyone below him to bother with study because they can always lean upon him to keep them out of trouble – and in any case 'the police usually come in and take over'.

Each reader must find his or her own niche and decide upon how far he or she wishes to progress with studies. Recommended reading includes –

Introduction to Criminal Law by Cross and Jones published by Butterworths.

The Law of Theft by J C Smith published by Butterworths.

The Decision to Prosecute by A F Wilson published by Butterworths.

Practical Security in Commerce and Industry by Oliver and Wilson published by Gower Press.

The Security Manual by the same authors and publishers.

Practical Fire Precautions by Underwood published by Gower Press.

Planning Fire Safety in Industry published by the Fire Protection Association who also publish much other useful informative literature about fire precautions and fire prevention in commerce and industry.

Fire Protection Association is at Aldermary House, Queen Street, London EC4N 1TJ.

Cost Effective Security by K G Wright published by McGraw Hill & Co.

The Investigation of Fraud by D Campbell published by Barry Rose Publishers.

Statutes and Publications referred to in this book and published by HMSO.

Retail Security – A Management Function by Byrne and Jones published by 20th Century Security Education Ltd.

Private study is very much a matter of self discipline. The student must be prepared to devote a given period to his studies with regularity. It is wrong to do three hours in one day and nothing for the next three days.

Far better would it be to study for one and a half hours on one day and half an hour on each of the next three days to revise and extend upon what he learned on the first day. Having started at that level it should not be long before the student is working four hours on day 1 and two hours on days 2, 3 and 4 of each week. If security work is to be studied with any degree of seriousness it requires 10 hours a week for six months or 15 hours a week for three months. Studying a subject does not only mean reading textbooks it means writing and re-writing passages from textbooks to commit the content of those passages to memory. It means the preparation of essays and written exercises to review particular aspects of the work and it also means reading about security in journals and periodicals wherever they can be obtained. It means getting involved in discussions and listening to lectures on security subjects of every kind regardless of whether those subjects affect our daily routines.

Broadening one's outlook on the subject means taking an interest in the other fellow's security problems, in retail, in building, in the marketing of equipment, in the storage and conveyance of inflammable liquids or gases under pressure and explosive substances. It means being prepared to lecture as well as to listen and it means involvement with policy planning, forward physical planning of premises, budgeting and other management functions as well as dabbling with daily duties.

In examination the objective should be to achieve the highest possible standard – to be a marksman – to aim at the bullseye all of the time – scoring a series of inners with answers will obtain a pass but scoring only outers will produce failure – one will only score at a high level by training oneself with effort and concentration.

The following advice was given to students who were preparing for an official examination:

'*Memory training*

A good deal of solid memorisation of information is necessary for this is the basis on which the process of recall can be applied. Apart from learning and memorising it is important that you understand what has been assimilated.

You will probably find you have more interest for some subjects than others. This is natural. The examination covers a wide field and it is essential that the mind, on a day-to-day basis of study, is trained to tackle those subjects for which it has little liking. This calls for self discipline.

Recall

The storage of word pictures in the memory banks is of no avail unless you develop your powers of recall. These can be assisted by inventing a triggering device which when activated opens the floodgates and releases information. This can be effected by learning a code word, each letter of which has some relevance to the separate essentials of a check-list.

Learning is best carried out with a colleague when memory recall can be tested by questions. At home a wife can be utilised for the purpose of asking questions and checking the answers.

At the commencement of studies learning by heart is best performed in short periods. It is advisable to limit study to one hour periods with a quarter hour break followed by another hour study period. This can gradually be increased when the mind is disciplined.

It has been found that when acts of reading and acts of recall alternate, ie when every reading is followed by an attempt to recall the items, the efficiency of learning and retention is enormously enhanced. This means that learning is best done by reading a paragraph or page or similar amount and immediately reciting it.

It has been found better to recite aloud than to recall in the head. The longer the passage that you set yourself to recall the better. If you find that you cannot recall the passage properly, read it again and then try another recall.

Tests have shown that when time is thus distributed between reading and recall fifty per cent more is remembered than when the same time is spent merely in reading the passage over and over.

Finally students should bear in mind that it is not sufficient purely to memorise Acts and Sections but must also understand what they are learning. Students should not hesitate to ask their tutors to further explain a point where necessary.'

The acquisition of knowledge is a task which everyone should tackle at a reasonably early stage in their lives because once mastered that knowledge can be of inestimable benefit for the remainder of one's working life and can open the way to career prospects and interests otherwise quite impracticable to attain.

Career Prospects in Industrial Security

The review of the growth of the security industry in Chapter 1 revealed that that growth took place very much upon a framework of ex-policemen and ex-firemen who brought to the industry knowledge and expertise acquired in other spheres of employment. Those stalwarts learned in the hard school of experience to adjust some of their former learning to meet company and employee points of view and the broad basis of their work still remains. The demand for supervisors and managers in this growth industry has outstripped the supply from those original sources and it has become necessary for the industry to promote from within and to draw upon other fields for its leadership.

In the past many security forces have been constructed of men from manual labour areas of industry whose skills have been overtaken by technology or who for health reasons have been taken off their normal work and have been glad enough to accept continued employment in a security department in lieu of redundancy or lay-off. Of latter years that scene has changed and there has been more emphasis upon direct recruitment to the lower ranks of security departments of fit men who possess intelligence, integrity and interest. Some of those men also possess a sense of leadership and have taken the first step or two up the ladder of promotion to become shift leaders, plant security supervisors or travelling inspectors and it is from those men that future chief security officers and security managers should be forthcoming.

In most of the larger companies there is today a promotional structure for security personnel. There are also openings in the smaller companies for men and women who have obtained groundwork experience and some promotion in the larger concerns to fill responsible security posts. In other concerns supervisory and senior security appointments have been filled by the recruitment of ex-Officers and NCOs from the three services as well as ex-members of the fire and police services.

Employers selecting a new head of a security department will look for an applicant who has a methodical approach to problems associated with the keeping of law and order allied to an ability to command and favourable consideration is likely to be given to an applicant who has held rank in a home or overseas civilian police force. Such an applicant knows the man in the street, has rubbed shoulders with him and understands his foibles, faults and failings – he is an applicant who not only has been trained but has had experience of handling subordinates, of making snap decisions and of exercising tact and discretion to the utmost limit. This applicant is almost readymade for the job but he does need to be given a period of training in business management where he can quickly learn that the primary objects of the business world are to create production, to give employment, to trade and to make a profit.

Commercial and industrial managers wishing to employ an ex-police officer for his experience and expertise must not expect him to accept the urgency of those objectives without assisting him in his transition from the sheltered non profit-making world of State employment. No longer does he wield authority under the protection of a warrant card or a baton and gone is the discipline code under which his life has been steered for so many years. Gone also is the respectful obeisance to those superior in rank to himself or from those in subordinate rank – he is now just plain mister and must get used to knowing people on first name terms.

His attitude towards offenders will probably need a re-assessment because he will find that the fellow whose collar he is about to feel suddenly acquires friends and sympathizers from unexpected quarters and there is a tolerance for offenders previously unknown to him. He must learn that people do not come to work to be disciplined and that the new and strange world of industrial relations must be examined and carefully assessed because it contains pitfalls for the unwary or unprepared ex-officer of whatever Service. Indeed the new security chief no less than the new gate guard does need training before he can be regarded as truly the man for the job.

It should be recognized that as the standard of new recruits improves and the amount of professional training in the security field increases in volume and quality there should emerge from the ranks of the profession a body of men well versed in the techniques of this occupation who will eventually compete, at least on equal terms, with former police and other professionals.

To merit those promotions the candidates must make themselves fit for the senior posts by study and by the acquisition of evidence that they are fit persons for such responsible positions and to this end mention has been made of the assistance to acquire professional status offered by the Industrial Police and Security Association and Institution of Industrial Security.

Some technical colleges and a few commercial companies now offer training for industrial and commercial security but they do not as yet prepare students for the professional status offered by the Institution of Industrial Security nor are they at present able to offer a comparable qualification.

There can be no better means of ensuring that the industry will continue to offer sound reliable security service to the world's business houses than for all security training to be undertaken by those within the industry qualified to instruct and to impart their own special brand of expert knowledge. It is only by the lower and middle ranks seeking and using that training that the industry will produce from within its security leaders of the future.

For the information of readers the National Secretary of the Industrial Police and Security Association is Mr L A Palmer, BEM, FIISec, of 7 Blue Waters Drive, Broadsands, Paignton, Devon. Mr Palmer is also Secretary to the Board of Governors of the Institution of Industrial Security. Correspondence should be accompanied by postage for reply when that correspondence relates to personal careers.

CHAPTER THIRTEEN

Fire Precautions – Legislation

There is a need at all levels of industrial security for knowledge of the law relating to fire precautions and of the arrangements for fire prevention and it is very timely that this book should be able to comment upon the most important and far-reaching change in fire precautions legislation affecting industry and commerce to have been introduced in this half-century. That change concerns the fire certification of factory, office, shop and railway premises and it came into effect on 1st January 1977. In order more easily to understand the current position it is advisable to review the position generally.

Whereas of latter years fire certification and fire precautions regulations for business premises have been imposed primarily by the Factories Act 1961 (hereinafter referred to as the 1961 Act) and the Offices, Shops and Railway Premises Act 1963 (hereinafter referred to as the 1963 Act) as from the above date the primary Act is the Fire Precautions Act 1971 (the 1971 Act). For five years the main provisions of this Act were applied only to hotels and boarding houses but those provisions are now broadly extended to premises 'used as places of work'.

Fire Precautions Act 1971

There is a tendency for modern legislation to be framed in such a way that clauses in the law, as laid down by various sections of an Act, may require a Secretary of State's designating order or commencement order or regulation to bring some or all of those sections into effect. This is so with the 1971 Act.

Section 1(1) of the Act requires that a fire certificate issued by the fire authority shall be required for all premises put to a use, or uses, designated in an Order made by the Secretary of State. Broadly those are uses of premises which involve members of the public being present there in any numbers, or in other circumstances in which they would be specially at risk in the event of fire.

By Section 1(2) a designating Order may be made for premises which fall within the following classes of use:

a) *use as sleeping accommodation;*

b) use as an institution providing treatment or care;
c) use for entertainment, recreation or instruction, or for purposes of any club, society or association;
d) use for teaching, training or research;
e) use for any purpose involving access to the premises by the public whether on payment or otherwise;
f) *use as a place of work*. (This clause was added to the original Act by Section 78 Health and Safety at Work Etc Act 1974.)

Designation of Premises used for sleeping accommodation

With effect from 21st February 1972 a designating Order (Statutory Instrument 1972 no 238) was made applying in England, Wales and Scotland the fire certificate provisions of the 1971 Act to hotels and boarding houses. The Order defined the designated premises as, 'hotels and boarding houses which provide sleeping accommodation for more than six persons, being guests or staff, and with sleeping accommodation above the first floor or below the ground floor'. A very useful commentary upon the intent and operation of the Act is contained in the following passage taken from the Home Office booklet *Guides to the Fire Precautions Act 1971 1. Hotels and Boarding Houses* first published by HMSO in 1972.

> 'When fire breaks out in a building in which people are present, the primary need is for those people to be able to escape safely and quickly even before the fire brigade arrives. Special experience and knowledge are needed to plan in advance what the escape routes should be and how they should be protected and equipped so that they will remain effective in an emergency for as long as is necessary. This is why the enforcement of this Act, which is designed to ensure the provision of adequate means of escape and related fire precautions in the premises within its scope, has been entrusted to the fire authorities. The main instrument of enforcement is the fire certificate which will be issued for particular premises only after the fire authority for the area is satisfied that the means of escape and other fire precautions are such as may be reasonably required in the circumstances of the case.
>
> The Act is concerned only with the protection of life in the event of fire. It cannot be applied to the protection of buildings and their contents, although it is likely that measures to protect life in the event of fire will also contribute to the protection of the building in which those measures are enforced.'

Very succinctly that passage sums up the mandatory obligation placed upon occupiers of premises. Clauses in fire certificates relate, amongst other matters, to adequate means of giving an alarm in the event of fire and to the adequacy of the means provided for fire fighting by persons on the premises but the main theme is for there to be provided safe routes of escape which will channel people out of buildings without confusion, reasonably protected from fire and smoke and with the capacity to give passage to the maximum numbers of persons expected to be in a building at any one time.

Designation of Premises used as places of work

With effect from 1st January 1977 a designating Order (Statutory Instrument 1976 no 2009) was made under Section 1 Fire Precautions Act 1971 and known as the Fire Precautions (Factories, Offices, Shops and Railway Premises) Order 1976. Article 3 of the Order applies the fire certificate provisions of the 1971 Act to premises defined by Article 2 of the Order as being:

> 'Factory premises' means premises constituting, or forming part of, a factory within the meaning of the Factories Act 1961, and premises to which Sections 123(1) and 124 of that Act (application to electrical stations and institutions, respectively) apply;
>
> 'Office premises, shop premises and railway premises' mean premises to which the Offices, Shops and Railway Premises Act 1963 applies and premises which are deemed to be such premises for the purposes of the Act, other than premises consisting of a covered market place wherein shop premises are aggregated;
>
> 'Covered market place' shall be contrued generally and not as limited to a place where a market is held by virtue of a grant from the Crown or of prescription or under statutory authority.

Article 4 of the Order is as follows:

1 Notwithstanding the foregoing provisions of this Order, a fire certificate shall not by virtue of Section 1 of the Act be required for factory premises, office premises, shop premises or railway premises being premises –

a) in which –

i not more than 20 persons are employed to work at any one time; or

ii not more than 10 persons are so employed elsewhere than on the ground floor,

unless any one of the conditions specified in paragraph (2) below applies to the premises; or

b) of a description specified in regulations which are from time to time made under the provisions of the Health and Safety at Work Etc Act 1974 and which provide for the issue of fire certificates by the Health and Safety Executive.

2 The conditions mentioned in paragraph (1)(a) above are –

a) that the premises are in a building which contains two or more of the premises described in Article 3 above and the aggregate of persons employed to work at any one time in all those premises exceeds twenty;
b) that the premises are in a building which contains two or more of the premises so described and in all those premises the aggregate of persons employed to work at any one time elsewhere than on the ground floor of the building exceeds ten;
c) that in the case of factory premises explosive or highly flammable materials (other than materials of such a kind and in such a quantity that the fire authority have determined that they do not constitute a serious additional risk to persons in the premises in case of fire) are stored or used in or under the premises.

It should be understood therefore that although all premises falling within the definition of factory premises, office, shop or railway premises outlined above are now designated for the purposes of the 1971 Act there is a clear demarcation between those requiring a fire certificate and those which are non-certified premises. Those requiring a fire certificate are premises in which more than 20 persons are employed either in a single occupancy or in the aggregate of a multiple occupancy, or where more than 10 persons are employed in like manner on floors other than the ground floor, or they are factory premises where explosives or highly flammable materials as prescribed are stored or used.

From the date of the Order the responsibility for the granting of fire certificates, for the general supervision of certificated premises and for the enforcement of fire precautions regulations in industrial and commercial premises was transferred from the Department of the Environment to the Home Office so that those responsibilities passed from the factory inspectorate into the hands of officers appointed by fire authorities. A fire authority is usually the County or County Borough Council or the City Council who act in direct liaison with the Fire Department of the Home Office.

This extension of the 1971 Act to 'places of work' had been envisaged by the Health and Safety at Work Etc 1974 in which Section 78 and Schedule 8 paved the way for this development and for the repeal of the general fire provisions of the 1961 and 1963 Acts. It was announced at the time of publication of the 1974 Act that the Goverment's policy would be to apply the Fire Precautions Act only to those places of work already covered by the 1961 and 1963 Acts and to ensure that the standard of fire safety was not reduced as a result of the transfer. The transfer of responsibility has therefore been a rationalization of the old law doing no more than to maintain the status quo, and therefore, in the main, only those premises formerly requiring a fire certificate under the 1961 and 1963 Acts are required to have a certificate under this extension of the 1971 Act.

Fire Certificates

By Section 40 of the 1961 Act a fire certificate was granted only if the fire authority was satisfied that the premises were provided with such means of escape in case of fire for the persons employed in the factory as may reasonably be required in the circumstances of the case. The certificate granted under this Section specified precisely and in detail the means of escape provided and contained particulars as to the maximum number of persons employed or proposed to be employed in the factory as a whole.

By Section 28 of the 1963 Act all premises to which the Act applied were required to be provided with such means of escape in case of fire for the persons employed to work therein as may be reasonably required in the circumstances of the case and also having regard to the number of persons, other than employed persons, who may reasonably be expected to be on the premises at any one time. A certificate granted under Section 29 of this Act specified the greatest number of persons who could safely be employed to work at any one time on the premises, the detail of the means of escape, and a statement that specified any special risks of the outbreak of fire inhere in the premises.

By Section 6(1) of the 1971 Act a fire certificate *will* specify:

a) the particular use or uses of the premises which it covers;
b) the means of escape in case of fire;
c) the means for securing that the means of escape can be safely and efficiently used at all times (including arrangements for emergency lighting, direction signs and smoke barriers);
d) the means for fire fighting for use by persons in the building;
e) the means for giving warning to other occupants of the building in case of fire.

By Section 6(2) of the 1971 Act a fire certificate *may* require the occupier to –

f) maintain the means of escape and keep them free from obstruction;
g) maintain other fire precautions specified in the certificate;
h) train members of the staff in raising an alarm, in fire fighting and in evacuation procedures, and to keep records of that training;

j) limit the number of persons who may be in the premises at any one time;
k) comply with any other fire precautions named in the certificate.

A fire certificate must be kept on the premises to which it relates and must be available for inspection by the fire authority at any one time.

The Act places an obligation upon the *occupier* of designated premises to apply for, and to be responsible for the compliance with the conditions of, a fire certificate. Where there is a multiple occupancy of a building, each occupier may be held to be responsible for his part of the premises but see the notes relating to the Fire Precautions Act 1971 (Modification) Regulations 1976 below.

Schedule 8 of the 1974 Act provided that all existing fire certificates issued under the 1961 and 1963 Acts continued in force after 1st January 1977 and will remain so until such time as they fall due for amendment or replacement or are revoked. The conditions of those certificates as set out above and as to the provision and maintenance of adequate means of escape in case of fire remain as they were under the former legislation but by authority of Section 78 of the 1974 Act *the following additional conditions are imposed by paragraph 2(3) of Schedule 8 that is to say:*

a) if the existing certificate was a 1961 Act certificate – all means of escape specified in the certificate shall be properly maintained and kept free from obstruction;

   while any person is within a factory for the purpose of employment or meals, the door of the factory, and of any room therein in which he is, and any doors which afford a means of exit for persons employed in the factory from any building or from any enclosure in which the factory is situated, shall not be locked or fastened in such a manner that they cannot be easily and immediately opened from the inside;

   any doors opening onto any staircase or corridor from any room in which more than ten persons are employed, and in the case of any factory constructed or converted for use as a factory after the end of June 1938, all other doors affording a means of exit from the factory for persons employed therein, shall, except in the case of sliding doors, be constructed to open outwards;

   in any factory constructed or converted for use as a factory before July 1938, in which more than ten persons are employed in the same building above the ground floor, any door which is not kept continuously open at the foot of a staircase affording a means of exit from the building, shall, except in the case of sliding doors, be constructed to open outwards;

every hoistway or liftway inside a building constructed after the end of June 1938, shall be completely enclosed with fire resisting materials, and all means of access to the hoist or lift shall be fitted with doors of fire resisting materials; except that any such hoistway or liftway which is not provided with a vent at the top shall at the top be enclosed only by some material easily broken by fire;

every window, door or other exit affording means of escape in case of fire or giving access thereto, other than the means of exit in ordinary use, shall be distinctively and conspicuously marked by a notice in letters of adequate size;

in every building which is, forms part of, or comprises a factory effective means capable of being operated without exposing any person to undue risk shall be provided and maintained for giving warning in case of fire, which shall be clearly audible throughout the building or, where the factory is part of the building in every part of the building which is used for the purpose of the factory;

the contents of any room in which persons are employed shall be so arranged or disposed that there is a free passage-way for all persons employed in the room to a means of escape in case of fire;

in any factory for which a fire certificate is required effective steps shall be taken to ensure that all the persons employed are familiar with the means of escape in case of fire and their use and with the routine to be followed in case of fire;

in every factory there shall be provided and maintained appropriate means for fighting fire which shall be so placed as to be readily available for use;

there shall be tested or examined at least once in every period of three months and whenever the fire authority so requires every means for giving warning in case of fire which is required to be provided for the premises;

there shall be entered in or attached to the general register the date of every test or examination carried out in pursuance of this regulation and particulars of any defects found and the date and particulars of any action taken to remedy any such defect;

if, after the grant of a fire certificate, it is proposed to increase materially the number of persons employed in the factory or in any part specified in the certificate, the occupier shall give notice in writing of the proposal to the fire authority.

b) if the existing certificate was a 1963 Act certificate –

all means of escape specified in a fire certificate shall be properly maintained and kept free from obstruction;

while a person employed to work in premises to which a fire certificate applies is in the premises for the purpose of doing his work or eating a meal, the doors of any doorway through which he might have to pass so as to get out of the premises shall not be locked or fastened that they cannot be immediately opened by him on his way out;

the contents of any room in premises to which a fire certificate applies, being a room wherein work is done by any of the persons employed to work on the premises, shall be so arranged or disposed as to afford to the persons who work in the room free passage-way to a means of escape in case of fire;

so long as a fire certificate in respect of any premises is in force, all exits affording, or giving access to, means of escape stated in the certificate to be relevant (other than exits in ordinary use) shall be distinctively and conspicuously marked by notices in letters of adequate size;

all premises to which a fire certificate applies shall be provided with effective means, capable of being operated without exposing any person to undue risk of giving warning in case of fire;

all means of giving warning in case of fire with which any premises are provided in pursuance of the preceding paragraph shall be tested or examined at least once in every period of three months and whenever so required by the fire authority;

effective steps shall be taken to ensure that all persons employed to work in any premises to which a fire certificate applies are familiar with the means of escape from the premises in case of fire and their use and with the routine to be followed in case of fire;

in all premises to which a fire certificate applies there shall be provided and maintained appropriate means for fighting fire which shall be so placed as to be readily available for use;

if, while a fire certificate is in force with respect to any premises, it is proposed to increase the number of persons employed to work therein at any one time above that stated in the certificate the occupier shall, before effect has begun to be given to the proposal, give to the fire authority notice of the proposal.

Other consequential legislation

With effect from the 1st January 1977 new legislation was introduced as follows:

a) The Fire Precautions (Non-Certificated Factory, Office, Shop and Railway Premises) Regulations 1976 made under the authority of Sections 12 and 37(3) of the 1971 Act

These Regulations relate to factories, shops, offices and railway premises in which persons are employed to work and in respect of which a fire certificate is *not* required by virtue of an Order made under Section 1 of the Act. These are the smaller premises where fewer persons are employed than in the premises requiring certification under the designation Order. The Regulations provide for an enforcement of fire safety measures in such premises with regard to the requirement to maintain the means of escape; the arrangement of outward opening exit doors and that an employee has the facility to open any door on his way out of the premises; the provision of enclosures providing fire resistance for thirty minutes around and in the access to every hoistway or liftway; the provision and maintenance of fire fighting equipment; the layout of workrooms and offices so as to afford a free passage-way to a means of escape in case of fire; and the provision of fire exit notices.

b) The Fire Precautions Act 1971 (Modifications) Regulations 1976 made under the authority of Section 80 Health and Safety at Work Etc Act 1974.

These Regulations modify parts of the 1971 Act in order to place obligations upon owners rather than occupiers in the cases of multiple occupancy premises. The 1971 Act in its original form specifically and deliberately placed responsibility upon each occupier of premises requiring fire certification but because the 1961 and 1963 Acts placed a responsibility for obtaining and complying with a fire certificate upon an owner of a building with a multiple occupancy a modification of the 1971 Act has been necessary in order to continue the authority of existing certificates – but strictly confined to its application to factories, offices, shops etc as defined under the designation Order.

c) The Fire Precautions Act 1971 (Commencement no 2) Order 1976.

This Order brought into operation certain additional Sections of the 1971 Act as required by this extension of its provisions to premises used as places of work.

d) The Fire Certificates (Special Premises) Regulations 1976 made by the Health and Safety Executive under authority of the Health and Safety at Work Etc Act 1974.

These Regulations provide for fire safety and certification alongside other safety provisions to be applied to premises such as mines, quarries, large petroleum establishments, chemical plants and similar premises with specially high danger risks.

e) The Fire Precautions (Application for Certificate) Regulations 1976 made under the authority of Section 5(1) Fire Precautions Act 1971.

These Regulations prescribes a new form of application for a fire certificate under the 1971 Act.

Special mention should be made here of the requirement under the designation Order 1976 no 2009 for there to be a fire certificate for a factory where explosives or highly flammable materials are used or stored irrespective of the number of persons employed. Such provisions did apply under the 1961 Act but factory inspectors in consultation with representatives of the fire authority were permitted to exercise discretion as to whether the quantity and nature of the materials stored or used justified certification and that flexibility of approach to this class of premises has been passed on to fire authority inspectors as the new enforcement officers.

Consequent upon the new legislation listed above the Department of the Environment has made regulations by which the whole of the fire provisions contained in Sections 40–52 inclusive of the 1961 Act and Sections 28–41 inclusive of the 1963 Act are repealed.

Home Office advisory publications

When in 1972 the designation Order was made bringing hotels and boarding houses under the provisions of the 1971 Act the Home Office issued a booklet *Guides to the Fire Precautions Act 1971 1. Hotels and Boarding Houses* as previously referred to in this chapter and they have now issued similar Guides relating the Fire Precautions Act 1971 to factories, and in a separate volume, to offices and shops. These Guides are most informative in describing the purpose and operation of the legislation to which they relate and should be read and used as a reference by every person having an interest in or responsibility for fire precautions arrangements in business premises. They contain such detail as the recommended minimum number of exit doors required from rooms of given sizes; the maximum distance of travel for any person within a room towards an exit which distance may be shorter in rooms of high fire risk such as kitchens or places where flammable materials are used; the maximum distance of travel from a room exit to a protected stairway or corridor leading to the outside of the premises; the requirement for each escape route to be screened by swing doors or similar means from the spread within a building of smoke and heat from fire; emergency lighting; ventilation; and recommended standards for the materials to be used for fire resisting wall, ceiling and floor construction and coverings. The booklets are obtainable from Government Bookshops and from good quality booksellers.

Effects of the new legislation

It should not be expected that there will be any immediate or dramatic effect created by this change of emphasis from the general supervision of factory, office, shop etc premises for fire safety by the factory inspectorate to the more specific supervision of such premises for fire precautions by a staff employed by each fire authority but the overall trend towards a greater awareness of fire risks, envisaged by the Roben's Committee on Industrial Safety, must progress as the organization of fire authorities grows.

The great majority of factories, offices and shops employing more than 20 persons at any one time or with more than 10 persons employed on floors other than the ground floor are usually reasonably well organized businesses controlled and managed by sensible people who take seriously their responsibilities for health, welfare and safety of the workforce and of the public frequenting their premises and they include in that responsibility an alert awareness of fire precautions. There are however badly organized concerns in that category where standards are not so high and where tighter enforcement controls are necessary to persuade proprietors of premises and reluctant employers that fire prevention arrangements are an essential part of running the business.

For an example of the way by which the new legislation can possibly effect improvement in fire prevention arrangements in such establishments let us consider the following clause in the law –

> '. . . there shall be provided and maintained appropriate means for fighting fire which shall be so placed as to be readily available for use.'

This direction was laid down in Section 51(1) 1961 Act and Section 38(1) 1963 Act and it is now embraced by paragraph 2(3) of Schedule 8 1974 Act to be one of the numerous additional conditions of a fire certificate under the 1971 Act.

Default in this aspect of fire precautions is commonplace. If fire extinguishers are shut in by storage or if appliances are tucked out of sight under a side desk or work bench or are allowed to hang upon the wall half empty or corroded it is regarded by the overworked employer to be no more than incidental to the running of his business – he will take the casual outlook that he is adequately insured, that he has done his bit by providing the equipment, and that it is for the staff to look after it. Factory inspectors may of course report adversely upon any shortcomings in fire precautions arrangements when inspecting premises to ensure the general health, welfare and safety of persons employed there but one knows how frequently those officers are steered away from areas where criticism can arise.

Under the 1961 and 1963 Acts penalty clauses were light and although penalties were raised considerably from 1st January 1975 under the provisions of the Health and Safety at Work Etc Act there has been no marked improvement over the past two years in the standards in premises to which the above critical comment refers.

Incorporating as a condition of a fire certificate the requirement for fire fighting equipment to be provided and maintained in position where it is readily available for use can, and undoubtedly will, bring about an improvement in premises affected by that casual outlook and more particularly when it becomes recognized that by Section 7(4) of the 1971 Act contravening any condition or requirement of a fire certificate is an offence for which the maximum penalty upon summary conviction is a fine of £400.

Some of the worst offenders in the matter of inadequate fire arrangements are to be found amongst those occupiers of factories, offices and shop premises which do not measure up to the category of fire certification and it will remain to be seen how effectively the fire authorities enforce compliance with Regulation 6 of the Fire Precautions (Non-Certificated Factory, Office, Shop and Railway Premises) Regulations 1976 which reads:

> '6(1) In all premises to which these Regulations apply there shall be provided and maintained appropriate means for fighting fire which shall be so placed as to be readily available for use.
>
> 6(2) In this Regulation "maintained" means maintained in an efficient state, in efficient working order and in good repair.'

When the inspectors appointed by fire authorities are organized to make more frequent visits to certificated and non-certificated premises as a matter of routine it is reasonably certain that casual and inconsiderate attitudes towards fire precautions will receive much closer scrutiny than heretofore. Additionally it is becoming increasingly noticeable that since the passing of the 1974 Act employed people have become more aware of their own and of their employer's responsibilities for health, welfare and safety and that awareness extends to a greater respect for safety equipment including fire fighting equipment. Failure on the part of the less responsible employers to provide, maintain and properly fix fire equipment is more likely in the future to result in that failure reaching the ears of the authorities than it might have done in the past and, of course, the authorities must follow up information of that nature.

Powers of inspectors appointed by fire authorities

Attention is called to the powers of inspectors appointed by fire authorities as contained in Sections 19, 20 and 40 of the 1971 Act and which may be summarized as follows:

Inspectors may take such action as is necessary for giving effect to the Act or regulations made under it. This includes power to enter at any reasonable times premises to which the Act applies, or seems to apply, as well as the rest of the building containing such premises. In the case of dwellings power of entry cannot be exercised as of a right unless 24 hours notice has been given to the occupier. Inspectors may make such enquiries as may be necessary to find out if the Act and regulations under the Act are being complied with and may require appropriate facilities and assistance to be given to them in the exercise of their powers. It is an offence to obstruct an inspector in the course of his duty or deliberately to give him false information.

An inspector visiting premises in connection with the enforcement of the Act is under a duty to produce evidence of his authority if asked to do so and this must be in the form of a duly authenticated document.

Over the past five years hotel and boarding house keepers have been compelled to meet the requirements of fire authorities representatives for expensive alterations to premises in order to comply with the standards set and recommended by the Home Office and which no traveller can have failed to notice in all classes of overnight accommodation in these islands. It would only be reasonable to suppose that, given time, similar stringent requirements will be applied to industrial and commercial premises to make compliance with fire certificate conditions more meaningful than has been the case in the past.

Other premises

In an appendix to the booklet about the application of the 1971 Act to hotels and boarding houses the Home Office lists other premises to which designation orders could be applied within the terms of Section 1(2) of the 1971 Act and these include hospitals, institutions and clinics providing treatment or care; places of entertainment, recreation or instruction including clubhouses; society headquarters or association meeting halls; schools and other teaching training or research establishments; and any other place to which the public are admitted whether on payment or otherwise and this includes museums, art galleries, places of historical or cultural interest, circus big tops and rivercraft moored or berthed and used for public entertainment.

The present position with regard to such premises is that for many of them fire precautions are included as conditions of licensing or are regulated by local Acts or Byelaws or deeds of covenant or are conditional for insurance purposes. It may be some considerable time before the fire prevention departments of local authorities are strong enough to cope with enforcement in all such premises but the long term plan to do so is clearly expressed in the 1971 Act and it must be accepted that it would be a very tidy package and a great improvement to have one standard of fire precautions for everyone and all premises and one enforcement body to exercise supervision over a pattern which is recognized by everyone controlling, using or employed in, any class of premises other than private dwellings.

By various Sections of the 1971 Act the following are excluded as potential designated premises – places of worship, dwelling houses with single occupancy, prisons, mental homes and premises used by the armed forces of the Crown, visiting forces and international headquarters of defence organizations.

An important provision of the 1971 Act is that contained in Section 10 which provides that 'whether or not premises within the scope of the Act are required to have a fire certificate the fire authority may apply to a court for an appropriate order if satisfied that the risk to persons in any premises in case of fire is so serious that the use of the premises in question ought to be prohibted or restricted'. This is in the nature of an emergency power so that the fire authority may take urgent action to deal with a serious and immediate fire risk in any premises whether designated or not.

New applications for fire certificates and variations in existing certificates

The provisions of the 1971 Act (Section 5) are very clear upon the procedure to be followed for obtaining the grant of a fire certificate and that procedure includes a survey of premises by the inspector appointed by the fire authority followed by the serving of a notice by the inspector upon the applicant informing him of work necessary before the grant of a fire certificate in order to bring up to standard the means of escape; the means for securing that the means of escape can be safely used (including emergency lighting – direction signs – smoke barriers etc); the means provided for fighting fire by persons in the building; and the means for raising an alarm within the premises in case of fire. It would appear to be a reasonable supposition that those obligations may be taken as the yardstick for the continuance of existing fire certificates after the first round of inspections by the new fire inspectorate.

Senior security officers are increasingly being called upon to advise management upon their legal standing with regard to fire, health, safety and welfare matters and those officers could give no better service than to consult with the fire authority for the area so that they fully and completely understand which rules, regulations and standards apply to their particular employer's premises and class of business whether certificated or not and without waiting for a critical inspection. The officers should then foster a wider interest in those standards and responsibilities by all levels of employees in the business so that compliance with the spirit as well as with the letter of the law becomes a matter of everyday routine to the mutual benefit of company and workers alike.

For future reading on this subject see –

Planning Programme for the Prevention and Control of Fire
Planning Fire Safety in Industry and Commerce

and numerous other advisory books and leaflets published by the Fire Protection Association Information and Publications Centre, Aldermary House, Queen Street, London EC4N 1TJ.

Fire Fighting in Factories, Part 10, of the Health and Welfare series published by HMSO Government Bookshop.

Fire Precautions Handbook by G W Underwood published by Gower Press Ltd.

Guides to the Fire Precautions Act 1971 for Hotels and Boarding Houses, Factories and Offices and Shops, published by HMSO Government Bookshop.

CHAPTER FOURTEEN

The Law and the Administration of Justice

1 Law is a system of rules arising from accepted custom and habit or enacted by authority and accepted by people for their guidance, protection and obedience. Those who do not obey or comply with the rules of law are said to act unlawfully or illegally and to the detriment of individuals or of the community as a whole.

Exemption from liability under the laws of this country is only accorded to the Sovereign, foreign sovereigns and foreign ambassadors residing here.

2 The law with which we are concerned is the law of England and Wales. Scotland retained her own system of private law after the union of 1707 but of recent years factors such as the sharing of a common legislature and a common court of final appeal have tended to make much Scots law resemble English law more closely than the continental law upon which Scots law was founded. The differences in administrative law in Scotland and England have become less marked as a result of the Crown Proceedings Act of 1947. The student who is concerned with Scottish law should bear in mind that marked differences do still exist and should take the time and the trouble to identify and to study those differences.

3 There are two categories of law known as Common law and Statute law.

Common Law From the earliest times certain rules of conduct have by custom and general approval become law. This law, known as Common law, is unwritten but accepted and approved by the courts. The term Common law was in use in the time of Edward I to describe that part of the law which was not statutory but common to the whole of England and in contrast to statute law, local custom and royal prerogative. Common law often means that part of English law which is unenacted, especially that contained in the decisions of the courts as opposed to acts of parliament. Because of the legislative supremacy of Parliament it follows that, where common law and statute law are in conflict, the latter prevails. A judge cannot refuse or neglect to apply an act of parliament on the ground that it is contrary to a fundamental principle of common law or that a development of common law has rendered the statute obsolete. On the other hand Parliament can, by legislation, abolish long established rules of common law, for example by Section 1(1) Criminal Law Act 1967 Parliament abolished all distinction between felonies and misdemeanours and, by Section 5(2) of the same act replaced the former common law offence of causing a public mischief by the statutory offence of 'wasteful employment of police'. Affray and conspiracy are two examples of common law offences which are still extant, all attempts to commit offences are themselves offences at common law except where an attempt is an offence by statute. The maximum penalties for attempts to commit crime are fixed by Section 7(2) Criminal Law Act 1967 as those to which the offender would be liable for a completed offence.

Statute law Statute law includes all laws made by direct order of the State and set out in acts of parliament (or statutes) made by the supreme authority in the country which is Parliament consisting of the Monarch, the House of Lords and the House of Commons. Every contemporary act of parliament commences with the following phrase:

> 'Be it enacted by the Queen's most Excellent Majesty, by and with the advice and consent of the Lords Spiritual and Temporal, in this present Parliament assembled, and by the authority of the same, as follows:–'

A statute is usually made by being first introduced as a 'Bill' to the House of Commons. This introduction is known as the 'First reading' and consists only of the reading of the title and the statement of the purpose of the Bill. This introduction is followed by a 'second reading' when the House has a general discussion followed by a debate on the principles of the Bill. The House may then accept or reject it. If accepted the Bill passes through the 'committee stage' when it is discussed in detail and any amendments are made. It then returns to the House of Commons for the 'third reading'. At this 'reading' the only amendments that are allowed are in the drafting of the Bill. The Bill is then sent to the House of Lords where the same procedure is applied. When the two Houses have agreed the Bill it is given the 'Royal Assent' by the Monarch and becomes Statute Law.

4 All law is again divided for administrative purposes into two main categories namely Civil law and Criminal law.

Civil law Some law is concerned only with the grievances of persons who feel that they have been wronged legally by the acts of others and who apply to the courts to have those wrongs put right. Courts dealing with such applications are known as Civil Courts where the parties involved in the action are known as the plaintiff and the defendant.

Examples of matters dealt with in civil courts, i.e. the Royal Courts of Justice and County Courts, are – actions to restrain others from continuing an existing nuisance or offensive trade; to seek the authority of the court to prevent the removal of property or persons abroad; to release from custody a person against whom no formal charge is being made; to require persons to pay their debts or to comply with conditions of an agreement which they have properly entered into under the law. Included in civil law actions are those concerned with the matrimonial laws and with all disputes concerning the possession of land and property and claims concerning wills and the break-up of business relationships.

A civil court is not so much concerned with punishment or exacting penalties as with arranging for grievances to be rectified or for wrongs to be put right or for compensation to be properly assessed but the court can and does exert pressure to those ends by imposing sanctions upon anyone who does not comply with its orders, directions and injunctions. Sanctions can go so far as the imposition of fines or committal to prison for contempt of court.

Some wrongful conduct may be both a civil and a criminal wrong and such conduct is known as a tort. All crimes are torts but not all torts are crimes. A tort is defined as a breach of a duty imposed by law. Any offender under the criminal law is liable to a civil action for damages taken by individuals hurt by the commission of an offence and examples may be found in assault, libel and injury caused to others by criminally bad driving. In such cases the injured party may claim in a civil court for compensation irrespective of whether or not a prosecution for a criminal wrong is being or has been taken in the criminal courts. Not all torts are crimes, for example Trespass is a civil tort for which a person aggrieved by the trespass may ask a civil court to have the trespass removed and may also seek compensation in that court for any loss suffered by the trespass. There are however no criminal proceedings to be taken for this tort because a mere act of trespass is not a punishable wrong. (Trespass is dealt with more fully in Chapter 17.)

Criminal law A criminal wrong is one by which a person offends against the rules of conduct applicable to all and thereby becomes liable to punishment by the State. Criminal wrongs may be prosecuted either by individuals or by the police laying an information for a warrant or summons at a magistrates court. A few civil wrongs such as non-payment of rates and failure to maintain one's family are dealt with in magistrates courts.

Crimes are dealt with in the Criminal Courts as described in later paragraphs and it is the operation of the criminal law and the criminal courts with which this Chapter is primarily concerned.

5 *Courts of Justice* There are several courts before which criminal offences may be heard and it is of some importance that security officers should understand how the law is administered. The courts are:

i Magistrate's Courts. All persons charged with criminal offences appear first at these courts where the proceedings are known as summary trials or as being dealt with summarily. These courts are also used for the preliminary hearing of the more serious cases which must go for trial before a Crown Court upon an indictment.

ii The Supreme Courts which include

a) The Crown Court which was formerly known as the Assize Court or Court of Quarter Sessions. Sittings of the Crown Court take place at many locations simultaneously and are usually held in County Towns and/or in the larger cities and towns. The proceedings of trial are known as cases dealt with on indictment.

b) The Court of Appeal where appeals against the findings of Crown Courts are heard.

iii The High Court of Parliament, ie The House of Lords which is composed of Lords of Appeal being lawyers of great emminence who are also members of the House of Lords. The Lords of Appeal deal with points of law of general public importance sent to them on appeal from decisions of the Supreme Courts.

iv Other courts dealing with facets of the criminal law are Coroner's Courts, Juvenile Courts and Domestic Proceedings Courts.

Security officers will have most reason to attend a Magistrate's Court where two or more Justices of the Peace sit together to hear the cases. JPs are usually persons not qualified in law assisted by the Clerk of the Court who is qualified. At Magistrate's Courts located in busy cities and towns there may be appointed a qualified and salaried Magistrate known as a Stipendiary Magistrate. The qualifications of a Stipendiary Magistrate must include seven years as a barrister or solicitor. A stipendiary magistrate sits on the bench alone but may be accompanied by JPs.

Clerks appointed to Magistrates Courts are nominated by the Magistrates' Court Committee and approved by the Home Office. Since 1949 it has been a requirement that Clerks so appointed should be barristers or solicitors of not less than five years standing and it is their duty to advise the Justices upon interpretations of law and upon questions of legal practice and procedure and the value and quality of the evidence presented to the Court. Justices are not obliged to accept the advice thus given.

All minor offences are dealt with summarily in a Magistrates Court. All major offences are first called before a Magistrates Court but whilst some may be dealt with summarily in that Court others must be committed for trial in the Crown Court upon indictment. Yet a third group may be dealt with summarily or on indictment according to the wishes of the accused person. That means that the accused has the right to elect to be tried either by a Magistrate or by a Judge and Jury. If he elects trial by the Magistrate he will be warned by the Clerk of the Court that should the case against him be proved and if the defendant's previous bad character and antecedent history are such that the Magistrate considers that his powers of punishment are insufficient to meet the circumstances of the case he may commit the accused to Crown Court for sentence.

The administration of justice in British Courts is quite complicated and the following notes cover only the broadest aspects of the subject.

The criminal jurisdiction of Magistrates is encased in the statutory law and falls mainly under two headings: (i) to conduct summary trials; (ii) to be an examining court for cases to be committed for trial.

In order to understand those two headings it is necessary to know that criminal wrongs are indictable, ie triable by judge and jury, unless the statute law specifies that they are summary offences. This means

a) that the most serious crimes and all common law crimes may only be dealt with on indictment;
b) that some lesser crimes may be dealt with on indictment or as summary offences according to the wishes of the accused or of the prosecutor; and
c) that minor offences may only be dealt with summarily.

Thus all persons charged with indictable and summary offences must first appear before a magistrates court.

Summary trial of summary offences

It is a fundamental principle of British criminal justice that all persons are presumed to be innocent of crime alleged against them until proven guilty. In summary trials the Clerk of the Court reads out the wording of the charge or summons and asks the person against whom the charge is made, and referred to as the defendant, whether he/she pleads 'guilty' or 'not guilty'. If the plea is 'guilty' the bench is informed of the facts of the case and the antecedent history and/or previous convictions of the defendant. Sentence is then pronounced and the case is concluded. If the plea is 'not guilty' the witnesses for the prosecution will give their evidence and may be crossexamined by or on behalf of the defendant. At the conclusion of that evidence the court will decide whether or not the defendant has a case to answer. If the court decides there is no case to answer (usually this is because the prosecution has failed to present sufficient evidence connecting the defendant with the offence charged) the court may dismiss the charge and set the defendant free. If the court decides that there is a case to answer the defendant must present his defence. The court may then either convict or dismiss the charge according to the evidence. If there is a decision to convict, the court may be told of the defendant's past character and antecedent history and may also hear a plea from the defendant for leniency and from any person in mitigation of sentence. The court will then pronounce sentence. If the court decides to dismiss the charge the defendant will be released without the court hearing anything of the defendant's character or past history.

Summary trial of indictable offences

If the case called is for an indictable offence which may be tried summarily the Clerk of the Court will read the wording of the charge or information to the defendant and will then ask him/her whether he/she wishes the case to be tried by the Magistrate(s) or by a Judge and Jury at Crown Court. He will at the same time inform the defendant that if he/she elects to be tried by the Magistrate(s) that in the event of the court finding the case proved but having regard to the defendant's previous convictions or antecedent history the court considers that the powers of punishment are insufficient to meet the circumstances of the case that the Magistrate(s) may commit the defendant to the Crown Court for sentence to be passed. If the accused elects to be tried by the Magistrate(s) he is then asked whether he/she pleads 'guilty' or 'not guilty'. Whatever the plea the proceedings continue as outlined under the previous sub-heading for summary trial of a summary offence. It should be noted that all theft is indictable and therefore the above formality is followed at Magistrates Courts for every case of theft coming before the court.

Committal proceedings for indictable offence

If the case called is for an indictable offence which may not be dealt with summarily or is for an indictable offence which the accused or the prosecution wish to have tried by Judge and Jury in the Crown Court the role of the Magistrate(s) is that of 'Examining Magistrate(s)' to be satisfied that there is a prima facie case to go for trial. At the outset of committal proceedings a decision is made which of two forms those proceedings will take.

Section 1 Criminal Justices Act 1967 authorises the Examining Magistrate(s) to commit for trial without consideration of the evidence provided that all of the evidence for the prosecution is before the court in the form of written statements, that all accused parties are legally represented, that copies of all statements have been exchanged by the legal representatives of the prosecution and the defence and that there is no objection to the content of those statements or plea by or on behalf of any of the defendants in the case that there is insufficient evidence contained in the statements for that defendant to be put on trial before a jury. Subject to those provisions the charge or charges are read over to the defendants and they are committed for trial without being asked to enter a plea and without any time being spent by the Magistrate(s) in reading the statements or listening to the statements being read in court.

The second form that committal proceedings may take is that which prevailed before the 1967 Act and which still will apply if any of the provisions of Section 1 of the Act is not fulfilled, for example, if an accused is not legally represented, or if there is an objection by or on behalf of an accused person to the content of a statement made by a prosecution witness. These proceedings will probably be quite lengthy because the prosecution may call all of its witnesses who will give evidence in chief and be subject to crossexamination. Then the defence may wish to call witnesses who can be crossexamined. The whole of the evidence will be recorded in longhand or in shorthand for transcription before the actual committal for trial takes place. Such recorded evidence is called making a deposition and the witness will have the record read over to him and be asked to sign each page of his deposition as being a true record of the evidence which he has given.

In effect these proceedings are a trial before a trial so that the Magistrate(s) are satisfied that there is a case upon which a jury should be asked to make a verdict. Being satisfied that there is such a case the Magistrate(s) may then commit the accused for trial and, under caution will ask him whether he has anything to say. Whatever the accused then says will be recorded and sent forward with the deposition to the Crown Court at which the trial will take place. Magistrates Courts may deal with minor cases completely at the first hearing and in petty cases for which there is a plea of guilty, this is frequently so but in any of its cases the court may adjourn or remand the hearing or part of the hearing to another date several days or even months hence.

Trials on indictment at the Crown Court

Cases which come before the Crown Court include –

i charges for serious crimes which may not be dealt with summarily;

ii charges for indictable offences against persons who have elected to be tried on indictment by Judge and Jury or for which the prosecution has asked for trial on indictment;

iii charges for indictable offences tried summarily but for which the lower court considered its powers of punishment to be inadequate and therefore have committed the accused to the higher court for sentence. These cases do not require the decision of a jury since the lower court has already found the case proved or have accepted a plea of guilty;

iv appeals against conviction and/or sentence by Magistrates Courts. Likewise these cases do not require the decision of a jury.

At Crown Courts, unlike the lower courts, once a case is called it is dealt with in entirety and lengthy cases may continue over several successive working days or even for several weeks. Trial cases are heard by a Judge sitting with a Jury, barristers usually represent the Crown as prosecutor. Accused persons may or may not be legally represented. If they are represented this will be by a barrister and in these days of free or assisted legal aid but very few accused persons do not take advantage of a barrister's services.

Rules of procedure for a criminal trial in Crown Courts are complex but can be of great interest and most educative and all men and women whose job in any way connects them with the law and its procedures would be well advised to spend a few days in the public gallery of the Crown Court to broaden their knowledge of life and of human involvements. Who knows but that before long they may themselves be deeply involved in one of those cases and at least an awareness of what is going on will be of the greatest benefit.

Appeals

A person convicted by a Magistrates Court may appeal to the Crown Court.

1 if he pleaded guilty – against sentence

2 if he did not plead guilty – against conviction or sentence.

There is no right of appeal to the Crown Court against the dismissal of a criminal case but application may be made for a case stated. Under the Magistrates Courts Act 1952 Sections 87 – 90 any person who was a party to any proceedings before a Magistrates Court or is aggrieved by the conviction, order, determination or other proceedings of the Court may question it on the ground that it is wrong on a point of law of some substance, or is in excess of jurisdiction, by asking the Court to state a case for the opinion of the High Court on the question of law or jurisdiction involved.

A person convicted of an offence on indictment (at the Crown Court) may appeal to the Court of Appeal against conviction

a) on any ground which involves a question of law alone, and

b) with the consent of the Court of Appeal on any ground which involves a question of fact alone, or a question of mixed law and fact or on any other ground which appears to the Court of Appeal to be a sufficient ground of appeal.

An appeal may go to the House of Lords at the request of either the prosecutor or the defendant from any decision of the Court of Appeal provided that leave to appeal has been given either by the Court of Appeal or the House of Lords. Such leave will not be granted unless it has been certified by the Court of Appeal that a point of law of such general public importance is involved that it ought to be considered by the House.

Aiders and Abettors

This Chapter has considered proceedings against offenders generally but attention is now called to the fact that offenders may not only consist of the actual perpetrators of crime but also to others who aid, assist, counsel or procure the commission of crime. Section 8 Aiders and Abettors Act 1861 enacts 'Whosoever aids counsels or procures the commission of an offence is liable to be indicted, tried and punished as a principal offender'.

Section 35 Magistrates Courts Act 1952 makes similar provision with regard to a person who aids, abets, counsels or procures the commissions by another of a summary offence. It matters not in either case whether the offence is arrestable or not.

Aiders and abettors may be prosecuted notwithstanding that the primary offenders are not known or have not been brought to trial if it can be proved

a) that the primary offence has been committed, and
b) that a person accused of aiding and abetting encouraged or assisted the commission of the primary offence and that he had the intent to aid its commission.

Mere presence at the scene of an offence without evidence of participation or encouragement is not enough to support a charge of aiding and abetting. Where one person moves into a position which bodily conceals the theft of goods by another then both parties may be charged as principals in the theft. If A and B each carry from a factory one component part of an article which is comprised of the two components and if both know that the article is being stolen by A they may both be charged as principals to the theft. If subsequently B claims that he did not know A was stealing it and carried his part upon the innocent belief that A had authority to take out the article it would be necessary for the investigation to have produced some sound evidence to negate that defence. It is doubtful whether the fact that B did not ask A whether or not he had a gate pass to take out the article would be sufficient evidence to convict B of theft or of aiding and abetting theft – some further evidence of mens rea (guilty knowledge) must be forthcoming to sustain a charge against B.

In a case in which a driver and his mate were jointly charged that they attempted by deception dishonestly to obtain property belonging to another with the intention of permanently depriving the other of it, contrary to common law, the case against the mate was dismissed on the grounds that he had taken no part in the attempted deception. There is good reason to believe that he would have been convicted had he been charged as an aider and abettor because evidence could have been given that he did know that the goods were to be obtained unlawfully.

In Chapter 19 the powers of arrest given by the law to a security officer are explained in detail and he is enjoined to use those powers with restraint and discretion – bearing in mind that by the apparently simple action of arresting a person without warrant he may set in motion a procedure which may end with an appeal to the House of Lords. With that sobering thought in mind no security officer is likely to use his powers thoughtlessly or without due consideration and care.

CHAPTER FIFTEEN

The Law of Theft

Whenever it is suggested that theft has taken place some person must examine the facts to establish whether the crime of theft has been committed. This may only be done by comparing the facts which are known or can be deduced by an enquiry, with the comparatively simple definition of theft as laid down in Section 1 Theft Act 1968. There is no other way of describing theft. It is therefore important that anyone whose duty it is to enquire into an allegation of theft shall thoroughly understand the definition which reads as follows:

THEFT ACT 1968 Section 1

1 A person is guilty of theft if he dishonestly appropriates property belonging to another with the intention of permanently depriving the other of it; and 'thief' and 'steal' shall be construed accordingly.

2 It is immaterial whether the appropriation is made with a view to gain, or is made for the thief's own benefit.

3 The five following sections of this Act shall have effect as regards the interpretation and operation of this section (and, except as otherwise provided by this Act, shall apply only for the purposes of this section).

In Sections 2, 3, 4, 5 and 6 of the Theft Act the legislators have endeavoured to clarify and interpret the various terms and expressions used in the Section 1 definition. Sections 2 to 6 do not create new offences they merely explain the basic definition of theft as laid down in Section 1 of the Act.

Property

The first point which an investigator needs to establish in his mind is whether the complaint he is examining is in respect of 'property' within the meaning of the Section 1 definition of theft. This is not all that difficult because Section 4 of the Act defines property as follows:

THEFT ACT 1968 Section 4

1 'Property' includes money and all other property, real or personal, including things in action and other intangible property.

2 A person cannot steal land or things forming part of land and severed from it by him, or by his directions, except in the following cases, that is to say –

a) when he is a trustee or personal representative, or is authorised by power of attorney, or as liquidator of a company, or otherwise, to sell or dispose of land belonging to another, and he appropriates the land or anything forming part of it by dealing with it in breach of the confidence reposed in him; or
b) when he is not in possession of the land and appropriates anything forming part of the land by severing it or causing it to be severed, or after is has been severed; or
c) when, being in possession of the land under a tenancy, he appropriates the whole or part of any fixture or structure let to be used with the land.

For purposes of this subsection 'land' does not include incorporeal hereditaments; 'tenancy' means a tenancy for years or any less period and includes an agreement for such a tenancy, but a person who after the end of a tenancy remains in possession as statutory tenant or otherwise is to be treated as having possession under the tenancy, and 'let' shall be construed accordingly.

3 A person who picks mushrooms growing wild on any land, or who picks flowers, fruit or foliage from a plant growing wild on any land, does not (although not in possession of the land) steal what he picks, unless he does it for reward or for sale or other commercial purpose.

For purposes of this subsection 'mushroom' includes any fungus and 'plant' includes any shrub or tree.

4 wild creatures, tamed or untamed, shall be regarded as property; but a person cannot steal a wild creature not tamed nor ordinarily kept in captivity, or the carcase of any such creature, unless either it has been reduced into possession by or on behalf of another person and possession of it has not since been lost or abandoned, or another person is in course of reducing it into possession.

Section 4(1) just about covers everything fixed or movable. Point one established is that money in any form is capable of being stolen. Point two is that so long as an item or article is real property or personal property it is capable of being stolen. *Real property* is that related to the realty as in 'real estate' and includes land and anything attaching to or growing on that land such as gardens, woodlands and hedgerows, buildings, fences and gates. So long as an ownership can be established the property can be the subject of theft. It is theft to dig up turf from a lawn or leaf mould from a wood, or to chop down a tree or to remove the bricks or stonework from a building or structure or to remove gravel from an open pit provided that in all cases the action is dishonest. *Personal property* means the contents of your house, your office or your car, it means the car, the corporation bus and the fleet of company owned lorries. Money of course is personal property. Pets and tamed or caged animals are personal property. It matters not whether the ownership is that of an individual or of a trust or corporate body or other group of persons. Personal property can be thought of as property which can be moved. Real property can be thought of as property which cannot normally be moved. When pieces of real property are severed they become the personal property of the person or body owning them as real property and he, or they, alone may authorise further sale, use or disposal or such pieces. There is an exception to this general principle given by Section 4(3) of the Act with regard to mushrooms growing wild and to flowers, fruit or foliage of a plant growing wild. Property thrown away or abandoned may remain the personal property of the owner of the land upon which the property is thrown or abandoned. Articles put into a council wastebin become the property of the council and to remove them without the authority of the council technically is theft but in such circumstances it would be a good defence for the appropriator to say that he believed the property to have been abandoned. Only where there is such abandonment as a bottle thrown into the sea does there cease to be ownership.

In order that there shall be as wide an interpretation as possible put upon the term 'property' Section 4(1) declares that the term shall include 'things in action and other intangible property' but the Act does not define those expressions. Lawyers tell us that the expressions refer to such things as bank accounts, a trademark, a debt, a copyright, a right to shoot over specified land or an hereditary right to fish certain waters or rivers, or a right of way. These are all abstract things and yet are capable of being stolen – they are all owned by someone.

In the matter of land ownership it is not theft to extend your boundary fence to encompass land owned by someone else but the aggrieved party may take action through a civil court to force you to cease your act of trespass and to replace your fence to its original position. However by Section 4(2) if a person is a trustee of land or has a power of attorney for the transfer, sale or other disposal of land belonging to another he will be stealing the land if he appropriates the land or any part of it and deals with it in breach of his trust. Furthermore although a tenant has the right to use fixtures on buildings let to him to be used with the land, and although he may prosecute as owner for the theft by another of any of those fixtures or from any of those buildings, he personally has no right to appropriate such things to his personal gain and to the detriment of his landlord. Section 4(3) provides a statutory defence to a charge of theft and says that 'a person who picks mushrooms growing wild on any land or who picks flowers, fruit or foliage from a plant growing wild on any land, does not (although not in possession of the land) steal what he picks, unless he does it for reward or for sale or other commercial purpose. Mushroom includes any fugus and plant includes any shrub or tree. A prosecution for theft of wild mushrooms or of wild flowers, fruit or foliage must be supported by *evidence* that the accused picked and appropriated those items with the intention of selling them or of otherwise making a profit from them and such evidence may not be forthcoming until an actual sale or transaction is completed.

Section 4(4) also gives a clear interpretation of the way in which Section 1 of the Act shall be applied to wild creatures and says 'wild creatures, tamed or untamed shall be regarded as property; but a person cannot steal a wild creature not tamed nor ordinarily kept in captivity, or the carcase of any such creature unless either it has been reduced into possession by or on behalf of another person and possession of it has not since been lost or abandoned, or another person is in course of reducing into possession'. It should be remembered that whilst this sub-section says that shooting or taking a wild rabbit or wild bird in flight or a fish in a river is not theft such shooting or taking may be poaching under the game laws which are not affected by the Theft Act.

To summarize this discussion upon the definition of 'property' for the purpose of Section 1 Theft Act it may fairly be said that with the exceptions

1 that land may only be stolen by a person breaking the trust placed upon him to dispose lawfully of land belonging to some other person; or

2 that wild mushrooms, wild flowers, wild fruit and the foliage of trees growing wild may not be the subject of theft unless taken for reward, sale or commercial purposes; or

3 that wild creatures not tamed nor ordinarily kept in captivity may not be the subject of theft unless they are stolen from some other person who has reduced or is reducing such animals to his own possession

all other property comes within the definition.

Belonging to Another

The next point to be considered by our investigator is to whom does the property belong? We have spoken of ownership but we must now turn our thoughts to the expression used in the Section 1 definition of theft and ask the question, is this property '*belonging to another*'? Section 5 of the Act contains five sub-sections which specify the circumstances in which property will be regarded in law to be 'belonging to another' as follows:

THEFT ACT 1968 Section 5

1 Property shall be regarded as belonging to any person having possession or control of it, or having in it any proprietary right or interest (not being an equitable interest arising only from an agreement to transfer or grant an interest).

2 Where property is subject to a trust, the persons to whom it belongs shall be regarded as including any person having a right to enforce the trust, and an intention to defeat the trust shall be regarded accordingly as an intention to deprive of the property any person having that right.

3 Where a person receives property from or on account of another, and is under an obligation to the other to retain and deal with that property or its proceeds in a particular way, the property or proceeds shall be regarded (as against him) as belonging to the other.

4 Where a person gets property by another's mistake, and is under an obligation to make restoration (in whole or in part) of the property or its proceeds or of the value thereof, then to the extent of that obligation the property or proceeds shall be regarded (as against him) as belonging to the person entitled to restoration, and an intention not to make restoration shall be regarded accordingly as an intention to deprive that person of the property or proceeds.

5 Property of a corporation sole shall be regarded as belonging to the corporation notwithstanding a vacancy in the corporation.

Under the provisions of Section 5(1) it will be seen that the person from whom property may be stolen, may be the shoe repairer, the watch repairer, the laundry or the garage owner carrying out work on customers goods. In such cases the goods may be said to 'belong to' the true owner *and* to the person temporarily in possession or control of them. The loser may be any one partner in a firm or a joint owner when partners or co-owners own or have possession or control of all property of the partnership or joint ownership. The loser may be a local authority or the ratepayers or a trustee or an auctioneer.

The tenant of whom mention has already been made is a person in possession and control of the furniture and fixtures rented to him and such items may be regarded as belonging to him so that in the event of theft by another party of any of those items the tenant may prosecute as the person having lawful possession and control of them. Yet he in turn may not dispose of any of those fixtures or the furniture to his own advantage because the landlord has not relinguished his control over them and therefore they still belong to him.

In effect what Section 5 says is that for the purpose of the Section 1 definition of theft property belongs –

i to any person or persons having ownership of it in the normal sense of the words
and at the same time
ii to any other person or persons having lawful possession or control of it
and at the same time
iii to any other person or persons having in the property a proprietary (owning) right or interest

and an act of theft may be complete if any of those persons is dishonestly deprived of property owned by them or assigned to them.

Appropriates

Our investigator having cleared his mind on the issues

1 that the subject of the complaint is property capable of being stolen, *and*

2 that that property 'belongs' to someone other than the accused

his next consideration must be whether the accused has 'appropriated' the property according to the meaning placed upon the word 'appropriates' by Section 3 of the Act which reads as follows:

THEFT ACT 1968 Section 3

1 Any assumption by a person of the rights of an owner amounts to an appropriation, and this includes, where he has come by the property (innocently or not) without stealing it, any later assumption of a right to it by keeping or dealing with it as owner.

2 Where property or a right or interest in property is or purports to be transferred for value to a person acting in good faith, no later assumption by him of rights which he believed himself to be acquiring shall, by reason of any defect in the transferor's title, amount to theft of the property.

The word 'appropriates' replaces the expression 'takes and carries away' used in the old definitions of larceny which are now obsolete and at the same time the word embraces a wider concept of conduct which amounts to stealing than did the old law. An assumption by a person of the rights of ownership includes all of the old ideas of taking and carrying away but also can include a transaction with someone else's property without touching or even seeing that property eg 'C' tells 'D' that he has an article for sale so that 'D' believing the story pays 'C' for that article and then removes the article from where it rests. 'C's' story was false and the article was not his to dispose of. 'C' has committed theft of the article although he has not laid hands upon it because he has appropriated it by assuming a right of an owner. Section 3(1) also tells us that the term 'appropriates' includes a case where a person has come by the property (innocently or not) without stealing it, and later assumes a right to keep it or to deal with it as owner. Hence something which is lent to another may be appropriated by the other if he does anything with it otherwise than return it to the person from whom he borrowed it. Another example would be when a person receives more than he is entitled to receive in a pay packet or in change over a shop counter – if he decides to keep the excess he is appropriating it and therefore stealing it.

Turning now to Section 3(2) if a person obtains property by a normal business transaction which he conducts in good faith and then he assumes his right to deal with or dispose of the property as its new owner he does not commit theft merely because the person from whom he acquired the goods had no right or authority to pass them on to him. The essence of this clause in the law is the genuine good faith in which the person receiving the goods acquired them. Once the new owner has been acquainted with the fact that the goods he has acquired are stolen property he may not dispose of them for profit because to do so would be to commit the offence under Section 15(1) Theft Act of – by deception dishonestly obtaining property (money) belonging to another with intent to permanently deprive the other of it.

It might be well to call the reader's attention here to the provision of Section 1(2) of the Act which says 'it is immaterial whether the appropriation is made with a view to gain, or is made for the thief's own benefit' and to the provision of Section 2(2) of the Act which says 'a person's appropriation of property belonging to another may be dishonest notwithstanding that he is willing to pay for the property'.

There is one further expression in the definition of theft which our investigator must consider and that is whether this appropriation of property belonging to another was made '*with the intention of permanently depriving the other of it*' and to explain the meaning of that expression Section 6 of the Act reads as follows:

THEFT ACT 1968 Section 6

1 A person appropriating property belonging to another without meaning the other permanently to lose the thing itself is nevertheless to be regarded as having the intention of permanently depriving the other of it if his intention is to treat the thing as his own to dispose of regardless of the other's rights; and a borrowing or lending of it may amount to so treating it if, but only if, the borrowing or lending is for a period and in circumstances making it equivalent to an outright taking or disposal.

2 Without prejudice to the generality of subsection (1) above, where a person having possession or control (lawfully or not) of property belonging to another, parts with the property under a condition as to its return which he may not be able to perform, this (if done for purpose of his own and without the other's authority) amounts to treating the property as his own to dispose of regardless of the other's rights.

Section 6 clearly illustrates the legislator's views of the kind of conduct which amounts to permanently depriving the owner (or person in possession or control) of property and rules out argument about whether or not neglecting to return a borrowed article amounts to theft. If the borrower has no intention of returning the article to its original owner or he has made some use of it which makes restoration of it in its original form impossible then he has committed theft.

Dishonestly

In order to satisfy himself on the points so far discussed, ie that property belonging to another has been appropriated with intent to permanently deprive that person of it, our investigator will have questioned the complainant or loser, the witnesses and other sources of information and from those persons or sources he will have learned that the appropriation has been carried out without the knowledge or consent of the person to whom the property 'belongs'. If there has been consent on the part of the loser then theft cannot have taken place. If that consent has been induced by some deception on the part of the taker then a crime against Section 15 Theft Act may have been committed and that is discussed in the next chapter. That apart it is to be assumed that the term 'dishonestly' used in the Section 1 definition of theft has its ordinarily accepted meaning and that where the appropriation is made without the knowledge and consent of the person to whom the property 'belongs' then that appropriation is a dishonest action. However Section 2 of the Act contains a proviso to that assumption and is as follows:

THEFT ACT 1968 Section 2

1 A person's appropriation of property belonging to another is not to be regarded as dishonest –

a) if he appropriates the property in the belief that he has in law the right to deprive the other of it, on behalf of himself or of a third person; or
b) if he appropriates the property in the belief that he would have the other's consent if the other knew of the appropriation and the circumstances of it; or
c) (except where the property came to him as trustee or personal representative) if he appropriates the property in the belief that the person to whom the property belongs cannot be discovered by taking reasonable steps.

2 A person's appropriation of property belonging to another may be dishonest notwithstanding that he is willing to pay for the property.

The last point to be established in the enquiry therefore is –

a) whether the accused is claiming a legal right to the property and if so whether he has a genuine *belief* in that right and is not just making an empty claim without justification. It matters not whether the legal right exists or not but so long as there are good grounds for the accused genuinely to believe that he has a legal right to the property he may not be convicted of theft. Cases in which there is a dispute as to right of possession fall within this sub-section and whilst it may be a matter for civil litigation to decide upon lawful ownership one of the parties may believe so strongly that he has a right in law to posses the article that he takes it into his own hands to deprive the other litigent of possession and thereby is accused of stealing it. If he is charged with theft it will not be the responsibility of the court of trial to decide upon lawful ownership but merely to confine its attention to a defence which claims that the accused appropriated the property in the *belief* that in law he had the right to deprive the other of it and the accused must be acquitted of the theft charge unless the prosecution can satisfy the court beyond reasonable doubt that the belief was *not* genuinely held; or
b) whether the accused is claiming that he sincerely believes that the owner would give him consent to have the property if he knew of the facts. Again it matters not whether the owner will or will not give his consent for the accused to have the property on this occasion. If there are good grounds for a person to believe that he would have the owner's consent he may be accused but he cannot be convicted

of theft. This is related closely to the practice of persons taking home perks from their work ie offcuts and surplus material the disposal of which the employer leaves to his work people and because of habit and custom known by the employer, employees assume and hold a genuine belief that consent will always be forthcoming. On the other hand if an employer has never given consent for offcuts and surplus material to be taken by employees such taking without his knowledge or consent amounts to theft because the employees have no genuine grounds for believing that consent would be forthcoming from their employer; or

c) whether the accused is claiming that he found the item abandoned and genuinely believed that the proper owner or loser would not be found even if he spent time upon an enquiry to trace him. Once again it is the genuine belief of the accused that matters and if that belief is sustained and supported by the facts of the finding then he may not be convicted of theft.

In effect Section 2 provides three instances of statutory defence to an accusation of theft. If the investigation reveals one or other of these excuses to be genuinely held there may be little point in pressing a charge. On the other hand although the investigation does not bring out such an explanation there will be many cases of theft which do go to Court in which the defence put forward in Court will be based upon one or other of the three clauses of Section 2. Whether or not they succeed will depend entirely upon the evidence in each case.

It would be quite wrong for the investigator to pointedly put the contents of Section 2 to an accused person or suspect and to invite him to comment upon his position with regard to them but when turning his attention to such an individual the investigator must have those issues in mind and should invite an explanation in such a way that a claim of innocence on one or other of the grounds set down in Section 2 will be naturally forthcoming if that is the true position.

The reader should refer back to these paragraphs again when studying the chapter dealing with the investigation of crime and it should then be appreciated that the reason for taking the various terms of this definition out of sequence in this chapter is in order that an investigation of an alleged theft shall follow the pattern of dealing firstly with complainants, losers, witnesses and others to obtain as much information as possible from them to establish that theft has taken place before asking the accused for an explanation.

The final enjoinder to our investigator must be to satisfy himself that *all* the elements of the definition are present before he can say 'This is a case of theft'.

The reader is now invited to test his understanding of this definition by assessing each of the following situations. Answers may be checked on the pages which follow the questions.

1 I take a packet of sugar off the shelf in a supermarket and put it into my basket instead of the supermarket basket and I do not offer to pay for it. As I go out of the shop an assistant challenges me and I say I am willing to pay for it and offer him the money.

2 My neighbour allows me to pick beans from his garden because he has more than he requires. He then goes on holiday and I pick beans in his absence.

3 My brother owns a bicycle and he and I both use it regularly. One day after an argument and to spite him I offer it to a friend for £18 but the friend does not accept.

4 My brother owns a bicycle and one day whilst I am using it with his consent a youth takes it from me and throws it down a cliff face.

5 Scouts are in camp on farmland. One boy digs up a root of blackberry from the hedge to take home. Others pick ripe blackberries to cook or eat.

6 I find a five pound note tucked between the pages of a book which I have borrowed from the library. I decide to keep the note.

7 A friend of mine is going abroad on holiday and asks me to collect his car from the airport car park. I go to pick up his car from the car park, and take the wrong one, because it is the same make and colour of my friend's car.

8 I pick up a purse near the shops, but there is no identification in it. I decide to keep it, though I know that the shopkeeper may know who has lost it.

9 'A' borrows a raincoat from a colleague in the office to keep him dry on his journey home one rainy evening. Ten months later the raincoat has not been returned and when the owner calls at 'A's' house to collect it he is told by 'A's' wife that she gave it away yesterday to a jumble sale because she was tired of asking her husband to take it back to the office.

10 A van driver finds that at the end of his daily round of delivering sausages to retailers he has four 4lb packs of sausages left on the van. All of his customers have signed for correct quantities and his delivery notes balanced with his load at the start of the day. May he keep the sausages as 'perks' of the job?

11 I find a signet ring on a shelf in the washroom at work and decide to keep it as there is nobody about to whom it could belong.

12 A company provides financial assistance for an employees' sports and social club to be maintained on land adjoining the company site. They appoint a steward to manage the club. During one evening of festivity in the clubhouse, some person gets behind the bar counter and takes five £5 notes from the till. From whom is the money stolen?

Answers

1 There are two factors here. The first is the theoretic commission of theft and the second the practical proof of theft. Theft of the sugar is complete at the moment the packet is picked up if the intent is not to pay for it but at that stage theft is not easy to prove because there is usually no evidence of intent. However, having passed the check-out point where I do not offer to pay for the sugar my intent is evident and I may be accused of theft. It is not necessary for me to leave the store completely to prove theft because I have passed the point at which an honest transaction could have taken place if that had been my intention.

2 Not theft. I hold a genuine belief that my neighbour would consent to my having the beans if he knew.

3 I commit an attempted theft by assuming a right of ownership to sell the bicycle and thus intend permanently to deprive my brother of it.

4 Theft is committed by the youth who regardless of ownership intends to deprive my brother and me of the bicycle and yet he does not want it for himself. At the time the bicycle was belonging to me as the person lawfully in possession and control of it.

5 Theft of the root which is severed from the realty (Section 4(2)(b). No theft of the fruit which was not picked for sale, reward or other commercial use.

6 This is theft. I came by the note innocently enough but I decide to appropriate it to my own use, although I know that by reporting the find to the librarian the owner of the note may be traced.

7 This is not theft because I believe the car to be my friend's car and that I have a legal right and authority to take it.

8 I have committed theft. If I believe that the shopkeeper would know whose it was, then I have not taken reasonable steps to trace the owner if I decide to keep the purse.

9 'A' can be said to have stolen the raincoat because his 'borrowing' amounted to an outright appropriation with no intention of returning it to the lender. 'A's' wife also has committed theft because the

property was never hers to dispose of and even though she gained nothing from the transaction.

10 No. These sausages are not a true perk for him because they are still the property of his employer who may yet receive a complaint of short delivery from a customer on the vanman's round.

11 This is theft because a signet ring is an identifiable thing and my belief that at the time of finding the ring the owner could not be found by taking reasonable steps cannot be justified or genuinely held to be true. If this signet ring is lost on company premises it becomes the personal property of the company until they decide that I may keep it because the true owner cannot be found. If it is the custom for 'found' property to be retained by the finder after reporting the matter, I must keep it in my possession for a reasonable time to allow time for a loser to come forward.

12 Persons to whom the money may be said to belong as it lays in the till are:

a) The members of the club who presumably share any profit made in the club by way of improved amenities or a Christmas party or in other ways, and
b) The company who have a proprietary interest in the financial position of the club, and
c) The club steward who has possession and control of the money in his till.

CHAPTER SIXTEEN

Other Crimes in Theft Act 1968

Robbery

The crime of robbery is one of the oldest and one of the most serious crimes known to English common law and has always described violent conduct committed for the purpose of stealing the personal goods of others. Through centuries robbers have been known by various terms such as footpads and highwaymen, and more recently as muggers and hijackers. Section 8(1) Theft Act 1968 has now replaced the old common law definitions of robbery by a new statutory definition which reads:

> 'A person is guilty of robbery if he steals, and immediately before or at the time of doing so, and in order to do so, he uses force on any person or puts or seeks to put any person in fear of being then and there subjected to force.'

By Section 8(2) the crime is punishable by imprisonment for life. The term 'violence' used in the old definition has been replaced by the term 'force' which implies a less aggressive use of physical strength and therefore widens the scope of the crime. The use of physical restraint in order to steal may now amount to robbery whereas formerly evidence of some personal attack or threat of personal hurt in order to steal was required to prove the crime.

The force used must be directed to the person of the victim and not against the property being stolen thus for a wage snatch to be committed by the snipping of a safety chain upon the wages bag without hurt being occasioned to the carrier is theft and will only become robbery if the carrier is forced to submit to the snipping of the chain by being held on the ground or is under threat of physical force to himself or if it has been necessary for the thieves to attack a bodyguard in order to get to the carrier and his bag. Pickpockets and handbag snatchers who steal from women who know or feel nothing until their property has gone are not robbers but are guilty of straightforward theft. A thief who tore a woman's ear in order to remove an earring was convicted of robbery but snatching a necklace by merely forcing the clasp was not held to be robbery.

To amount to robbery there must be theft and all of the argument and proof that theft has been committed will apply before the Court may consider the aspects of the use of force in order to convict for robbery. For this reason Section 8(2) Theft Act encompasses the alternative crime of assault with intent to rob which may be charged against an accused who attacks someone with theft by the use of force in mind but who fails to complete his theft. This crime is also punishable by life imprisonment.

It is important to note that in robbery the force used must be 'immediately before or at the time of stealing AND IN ORDER TO DO SO'. If theft takes place without the use of force any subsequent use of force to escape will not make the crime robbery – but should result in charges of theft plus an appropriate charge under the Offences Against the Person Act 1861 for the subsequent assault.

Burglary

By Section 9 Theft Act 1968

1 A person is guilty of burglary if:

a) he enters any building or part of a building as a trespasser and with intent to commit any such offence as is mentioned in sub-section (2) below; or
b) having entered any building or part of a building as a trespasser he steals or attempts to steal anything in the building or that part of it or inflicts or attempts to inflict on any person therein any grievous bodily harm.

2 The offences referred to in sub-section (1)(a) above are offences of stealing anything in the building or part of a building in question, of inflicting on any person therein any grievous bodily harm or raping any woman therein, and of doing unlawful damage to the building or anything therein.

3 References in sub-sections (1) and (2) above to a building shall apply also to an inhabited vehicle or vessel, and shall apply to any such vehicle or vessel at times when the person having a habitation in it is not there as well as at times when he is.

4 A person guilty of burglary shall on conviction on indictment be liable to imprisonment for a term not exceeding fourteen years.

Students of the criminal law before the passing of Theft Act in 1968 will note that there is now no distinction between night and day; that the term burglary no longer relates to dwelling houses alone but to all buildings; and that the idea of 'breaking into' or 'breaking out of' premises has gone and it merely has to be proved that the accused was a trespasser in any kind of building or in any particular part of a building and either

a) that he entered the building or part of the building with an intent to commit one or more of the crimes listed in Section 9(2); *or*
b) that having entered as a trespasser with or without any such intent he nevertheless does commit theft or attempts to commit theft or does cause to some person therein some grievous bodily harm or does attempt to cause grievous bodily harm therein.

For example, a tramp who enters a derelict building to sleep but who breaks up a floor board to burn on a fire to keep himself warm becomes a burglar under Section 9(1)b) by reason of his theft of the board even though theft was not in his mind when he entered the building. The tramp who finds another tramp occupying his 'warm corner' and so attacks him to cause him to leave will be a burglar under Section 9(1)(b) if his attack amounts to grievous bodily harm or an attempted grievous bodily harm even though he did not have violence in mind when he entered the building.

It should be noted that the Act goes so far as to take heed of those who trespass into parts of premises to which they have no right of access thus we are told that a production employee who trespasses into the personnel department during a period when that department is unoccupied for the purpose of finding confidential information will not be committing a crime but if the employee decides to remove a document from that department or uses a sheet of paper found therein upon which to copy information and then removes the sheet from the department he will have committed theft Section 1 and burglary Section 9(1)(b).

Apropos the first illustration the tramp who entered merely to sleep does not become a burglar if his later conduct amounts only to committing damage to the building or its contents and in the second illustration the tramp's attack upon the other would not be burglary if his attack amounts to rape when the other tramp in his warm corner turns out to be a woman. Intent at the time of entering the building is the vital factor in these cases.

The illustrations given are intended to impress the reader with the significance and importance of properly understanding the definitions because correct action upon finding an offender so much depends upon a quick and accurate assessment of the facts. The tramp will be arrested regardless of whether he is subsequently to be charged with burglary, theft, criminal damage or rape because all are arrestable offences. He would only be in the clear if he just crept into the building to sleep and crept out again. But a quite different proposition is the employee snooping about for information in the personnel department because if he finds what he is looking for he can cause dissention amongst his fellow workers, or undermine an agreement with the company or perhaps indulge in a little blackmail and if he is not caught doing damage or stealing something besides information there is very little that anyone can do about it.

On the other hand for a security officer to recognize that that piece of paper in the intruder's hand belongs to the department and is vital evidence for charges of theft and burglary is most important and can be the difference between the offender going scot free to stir up such trouble as he pleases, or for him to be charged with crime and lawfully sacked from his employment for wilful misconduct.

The expression 'enters a building' will be subject to all former case law upon what amounts in law to an entry and students are reminded that the expression includes the entry of an instrument used not merely for the purpose of breaking in, but as an extension of the body in effecting the particular crime, eg the 'entry' of a jemmy used to smash a window would not be sufficient, but a rod inserted through a hole in a jeweller's window to remove a necklace, or the barrel of a shotgun pushed through a doorway with intent to cause grievous bodily harm would be.

The expression 'building' is not defined in the Act and it is to be assumed that this infers a structure of some permanence with four walls and a roof. A tent would not be a building within this Section nor would a piece of apparatus such as a street telephone kiosk.

The subject of trespass and the bearing which that term has upon the crime of burglary is dealt with fully in Chapter 17 which should be read in conjunction with this subject.

Burglary is an arrestable offence but so also are theft, rape, grievous bodily harm, and criminal damage whether to premises or to contents and if there is evidence that *any* of those crimes has been committed the civilian's power of arrest may be used to arrest (detain) any person suspected on reasonable grounds of having committed that crime. It can be sorted out later whether burglary may be proved. If there is no evidence of the actual commission of theft, rape, grievous bodily harm or criminal damage a decision must be made as to whether there is evidence to support an accusation under Section 9 (1)(a) of entering as a trespasser with intent etc and the big question is why did he enter?

In Chapter 18 advice is given about the first steps in an investigation of crime and in the interrogation of intruders and that advice needs to be closely followed to assess whether you have caught a burglar. It is not proposed to go over that advice again here.

To summarize. A charge of burglary under Section 9(1)(a) may be made if the circumstances are such, and the evidence is available, to prove the intent of the intruder when he entered the premises or that part of the premises wherein he was trespassing. A charge of burglary under Section 9(1)(b) may be made without proving intent at the time of entering the premises if there is evidence that during the trespass, and regardless of the reason for which he is trespassing upon the premises or upon that part of the premises, he stole property or attempted to steal property or caused or attempted to cause to anyone there any grievous bodily harm.

Section 10 Theft Act 1968 refers to the crime of Aggravated Burglary as follows:

> A person is guilty of aggravated burglary if he commits any burglary and at the time has with him any firearm or imitation firearm, any weapon of offence, or any explosive.

Burglary must be proved plus the fact that the accused 'had with him' a firearm, weapon of offence, or explosive which are defined in the Section as follows:

> Firearm includes an airgun or air pistol, and imitation firearm means anything which has the appearance of being a firearm, whether capable of being discharged or not; and
>
> Weapon of offence means any article made or adapted for use for causing injury to or incapacitating a person, or intended by the person having it with him for such use; and
>
> Explosive means any article manufactured for the purpose of producing a practical effect by explosion, or intended by the person having it with him for that purpose.

The penalty for aggravated burglary is life imprisonment. Aggravated burglary should only be regarded as a graphic extension of the crime of burglary.

Taking a Motor Vehicle or other Conveyance without Authority

Section 12 Theft Act 1968 reads as follows:

1 Subject to sub-sections (5) and (6) below, a person shall be guilty of an offence if, without having the consent of the owner or other lawful authority, he takes any conveyance for his own or another's use or, knowing that any conveyance has been taken without such authority, drives it or allows himself to be carried in or on it.

2 A person guilty of an offence under sub-section (1) above shall on conviction on indictment be liable to imprisonment for a term not exceeding three years.

3 Offences under sub-section (1) above and attempts to commit them shall be deemed for all purposes to be arrestable offences within the meaning of Section 2 of the Criminal Law Act 1967.

4 If on the trial of an indictment for theft the jury are not satisfied that the accused committed theft, but it is proved that the accused

committed an offence under sub-section (1) above, the jury may find him guilty of the offence under sub-section (1).

5 Sub-section (1) above shall not apply in relation to pedal cycles; but, subject to sub-section (6) below, a person who, without having the consent of the owner or other lawful authority, takes a pedal cycle for his own or another's use, or rides a pedal cycle knowing it to have been taken without such authority, shall on summary conviction be liable to a fine not exceeding fifty pounds.

6 A person does not commit an offence under this section by anything done in the belief that he has lawful authority to do it or that he would have the owner's consent if the owner knew of his doing it and the circumstances of it.

7 For purposes of this section –

a) 'conveyance' means any conveyance constructed or adapted for the carriage of a person or persons whether by land, water or air, except that it does not include a conveyance constructed or adapted for use only under the control of a person not carried in or on it, and 'drive' shall be construed accordingly; and

b) 'owner', in relation to a conveyance which is the subject of a hiring agreement or hire-purchase agreement, means the person in possession of the conveyance under that agreement.

The offences in sub-section (1) can best be grouped as follows:

i without the consent of the owner or other lawful authority taking a conveyance for his own use or the use of another;

ii knowing that such an offence has been committed drives the conveyance or allows himself to be carried in or on it.

The distinction between offences under this Section and straightforward theft of a vehicle is the absence of evidence to 'permanently deprive the owner' of the vehicle as for example after use and abandonment but parties driving and being carried *before* abandonment may find themselves charged with theft and it is for them to convince the court of their ultimate intention to return the vehicle to the owner or to abandon it. In such a case sub-section (4) will apply. This is one of the few crimes which are arrestable offences even though the maximum penalty on first conviction is less than five years.

This offence had its origin in the road traffic laws and its transfer into the Theft Act illustrates that the law regards the offence to be more closely related to the deprivation of an owner of his property, albeit only temporarily, than to any technical breach of the safety rules of driving by joyriding.

When the expression 'takes' was used in the old definition of larceny it was generally accepted that a person 'took' an article if he acquired possession of it. Relating this to the Theft Act concept of 'appropriation' it may be reasoned that a 'taking' would occur whenever a person exercises a control over the use of a conveyance without the consent of the owner or other lawful authority. Under the road traffic laws the taking had to be accompanied by driving but this is not in Section 12 Theft Act and so rolling downhill, pushing or bodily transporting a conveyance would now be a taking as also would floating away a boat. Case law has established that a lorry driver who merely deviated from his route for a private purpose was not guilty under the Road Traffic Act of 'taking and driving away' his lorry but another driver who parked his employer's vehicle outside his house after work journeys were concluded and then subsequently drove the vehicle on a journey for his private purpose was guilty of 'taking and driving' under the same Act. Such rulings must be regarded as appropriate to the offence of unlawful taking a conveyance under Section 12 Theft Act.

The expression 'conveyance' as defined in sub-section (7)(a) does not include a pedestrian operated vehicle and although it does include a mechanical horse upon which the driver rides it does not include the trailer body drawn by it since that part of the vehicle is not made or adapted for the conveyance of a person.

'Consent of the owner' must be true consent and does not include consent obtained by a false or fraudulent story.

Attention is called to sub-section (6) which provides a statutory defence in circumstances in which the accused held an honest belief that he did have lawful authority to take the conveyance or held an honest belief that he would have the owner's consent if the owner knew of the circumstances under which he took the conveyance.

Once again we meet the proviso in the law as given in Section 2 Theft Act 1968 and in Section 5 Criminal Damage Act 1971 when an apparent criminal action is excused if there is a genuine belief held by an accused person that he thought he was within his rights. Such explanations need to be very carefully examined before acceptance by an investigator but they must not be ignored.

Sub-section (1) provides that any passenger carried on a 'taken conveyance' shall be guilty as a principal offender if it can be proved that he knew that it was unlawfully taken. Care should be taken to note anything said or done by passengers to illustrate their knowledge so that they do not escape the consequences of their action by posing as innocent parties.

Sub-section (5) creates the separate offence of the unlawful taking of a pedal cycle; there is no power of arrest; the maximum penalty is a fine of £50; the proviso contained in sub-section (6) does apply in these cases – if one person takes it and another rides it knowing that it was unlawfully taken they commit separate offences under the sub-section and are each liable to the maximum fine.

Finally, sub-section (7)(b) defines the 'owner' as including a person who normally possesses a conveyance under a hiring agreement or a hire purchase agreement.

Nowhere in the Section is there a limitation upon the place from which a 'taking' may occur. This offence was originally a traffic act offence and, since it has now been taken out of the road traffic laws and placed into the Theft Act, it is to be assumed that a 'taking' inside a works car park would be complete without the conveyance leaving the car park ie joyriding in the park on the vehicle of another.

It would indeed appear to include as an offence under the Act the 'taking' of a fork lift truck inside works premises and to 'joyride' it around a yard or warehouse without consent. A few prosecutions should stop some of that dangerous horseplay which takes place on occasion, and the posting of notices to that effect might be worthy of consideration.

Abstracting Electricity

Section 13 Theft Act 1968 provides that a person who dishonestly uses without due authority, or dishonestly causes to be wasted or diverted, any electricity shall on conviction on indictment be liable to imprisonment for a term not exceeding five years. This is an arrestable offence.

The source of the electricity consumed is not defined and may be a public supply or a private supply such as that generated inside works premises or it may be from a storage battery.

Employees who make unauthorized telephone calls for private purposes over company telephone systems or who use a power supply in a workshop for private work which is not authorized commit an offence under Section 13. It might be advisable in some locations to post notices to this effect.

Fraud in Industry and Commerce

1 Obtaining property by deception

Section 15 Theft Act 1968 provides:

1 A person who by any deception dishonestly obtains property belonging to another, with the intention of permanently depriving the other of it, shall on conviction on indictment be liable to imprisonment for a term not exceeding ten years.

2 For purposes of this section a person is to be treated as obtaining property if he obtains ownership, possession or control of it, and 'obtain' includes obtaining for another or enabling another to obtain or to retain.

3 Section 6 Theft Act shall apply for purposes of this section with the necessary adaption of the reference to appropriating as it applies for purposes of section 1.

4 For purposes of this section 'deception' means any deception (whether deliberate or reckless) by words or conduct as to fact or as to law, including a deception as to the present intentions of the person using the deception or any other person.

Note the similarity of wording between the offence under Section 15(1) and theft under Section 1(2) of the Act. There must be an obtaining of property in place of an appropriation of property and Section 15(2) defines what is meant by obtaining.

Section 34(1) Theft Act provides that the Section 4(1) definition of 'property' and Section 5(1) definition of 'belonging to another' shall apply generally for the purposes of the Act as they apply to Section 1.

Section 15(3) applies the definition of Section 6 'with intent permanently to deprive the other of it' to this Section with the minor adjustment of exchanging the term 'obtaining' for 'appropriation'. This leaves the expression 'by any deception dishonestly' needing clarification.

Section 15(4) provides the meaning of 'by any deception' whilst the term 'dishonestly' has a normally accepted meaning of the word without the special exceptions given to it for the purpose of Section 1(1) of the Act by Section 2. Dishonesty in an action implies an intention to defraud.

Under Section 15 it is necessary to prove that the property was obtained as a direct result of a deception. Obtaining postage stamps from vending machines by using metal discs in place of coins is theft by Section 1 Theft Act because the discs act as a false key or jemmy to get at the stamps and the Post Office does not lose its stamps by deception. If however D enters a sub-post office and tells the sub-postmaster an untruthful story that on the previous evening he had put money into the stamp machine on the outside of the premises and that no stamps had come out and if the sub-postmaster accepts that story and hands over to D stamps or the amount in cash that D says that he put into the machine there would be an offence of obtaining property (stamps or money) by deception contrary to Section 15(1).

Generally theft is complete when a person parts with possession of his property without his knowledge and/or consent but it is cheating, ie an offence under Section 15(1), when a person knowingly parts with possession of his property having been deceived into doing so by a false, fraudulent or misleading story.

2 Obtaining pecuniary advantage by deception

Section 16 Theft Act 1968 provides:

1 A person who by any deception dishonestly obtains for himself or another any pecuniary advantage shall on conviction on indictment be liable to imprisonment for a term not exceeding five years.

2 The cases in which a pecuniary advantage within the meaning of this Section is to be regarded as obtained for a person are cases where –

a) any debt or charge for which he makes himself liable or is or may become liable (including one not legally enforceable) is reduced or in whole or in part evaded or deferred; or
b) he is allowed to borrow by way of overdraft, or to take out any policy of insurance or annuity contract, or obtains an improvement of the terms on which he is allowed to do so; or
c) he is given the opportunity to earn remuneration or greater remuneration in an office or employment, or to win money by betting.

3 For purposes of this section 'deception' has the same meaning as in Section 15 of this Act.

The offence is comprised of conduct by which a financial advantage is obtained or by which an opportunity for financial reward or credit is obtained *prior* to the acquisition of property in the form of money or other valuable security. In this respect Section 16 offences differ from those under Section 15 wherein actual acquisition of property (in any form) must be proved.

There is only one offence ie that under Section 16(1) but it is expected that particulars of an offence should show whether it is based upon Section 16(2)(a) or (b) or (c).

Under Section 16(2)(a) the liability for a debt or charge may relate to

i a debt or charge for which the offender *makes himself liable* ii a debt or charge for which the offender *is liable* iii a debt or charge for which the offender *may become liable*	which is reduced or is evaded or is deferred

and if such reduction, evasion or deferment is occasioned by deception on the part of the offender he will be guilty of an offence under this Section. Obviously cases of partaking of a meal in a restaurant without the means to pay for it which formerly would have been an offence under the Debtors Act 1869 are now cases under Section 16(1) qualified by 16(2)(a) and there may be similar cases in which a person uses a service or obtains the use of equipment when there is no requirement to pay in advance but for which he has no means of paying when demand is made for payment and a typical example is when a person uses a taxi service upon the assumption that he will pay at the end of the journey but has no money or means by which to do so.

An example of a case under Section 16(1) qualified by 16(2)(a) would be one in which a person gave false information as to his means or income upon an income tax form or a form relative to some social service benefit so that any subsequent charge for which he may become liable is reduced or evaded as a result of the false information given.

Section 16(2)(b) means just what it says.

Section 16(2)(c) covers the offence in which a person obtains employment by falsely stating that he has a qualification which he does not have and which is essential for acceptance for the job. Any payment for the job thereafter received would be his financial advantage obtained by deception. Similarly if (A) puts in a false reference which results in (B) getting a job he otherwise would not have been given (A) commits an offence under 16(1) qualified by 16(2)(c).

There are so many complicated interpretations upon the wording of Section 16 that Lord Justice Edmund Davies has called the Section a judicial nightmare and expressed the hope that it would soon be replaced by a simpler provision.

3 False accounting

Section 17 Theft Act 1968 provides:

1 Where a person dishonestly, with a view to gain for himself or another or with intent to cause loss to another –

a) destroys, defaces, conceals or falsifies any account or any record or document made or required for any accounting purposes; or
b) in furnishing information for any purpose produces or makes use of any account, or any such record or document as aforesaid, which to his knowledge is or may be misleading, false or deceptive in a material particular;

he shall, on conviction on indictment, be liable to imprisonment for a term not exceeding seven years.

2 For purposes of this Section a person who makes or concurs in making in an account or other document an entry which is or may be misleading, false or deceptive in a material particular, or who omits or concurs in omitting a material particular from an account or other document, is to be treated as falsifying the account or document.

This Section is extremely broad and very valuable for challenging the action of persons who have altered figures or made other changes to accounting documents which will eventually be to their financial advantage or to the benefit of another but before that advantage or benefit materializes.

Examples of this offence would be if employees make false entries upon time sheets as to the hours they have worked or they alter any such entry so that they receive payment for more hours than actually worked; or when one employee clocks an attendance card for another who at that time is absent from the premises; or when more time than has been worked is booked for a class of job carrying a higher rate of pay. The offence also arises when, in order to gain extra benefit from a tonnage or bonus scheme, an employee alters figures on delivery or returns notes or on weighbridge tickets or similar documents.

It is as much an offence under Section 17 to omit to make a necessary entry as it is to make a false one and it is a complete offence to present or use a misleading document, or one containing false information, whether or not it is accepted as genuine.

The reader's attention is called to the end of Chapter 18 where reference is made to the special nature of investigations into fraud cases.

Blackmail

Section 21 Theft Act 1968 provides;

1 A person is guilty of blackmail if, with a view to gain for himself or another or with intent to cause loss to another, he makes any unwarranted demand with menaces; and for this purpose a demand with menaces is unwarranted unless the person making it does so in the belief

a) that he has reasonable grounds for making the demand; and
b) that the use of the menaces is a proper means of reinforcing the demand.

2 The nature of the act or omission demanded is immaterial, and it is also immaterial whether the menaces relate to action to be taken by the person making the demand.

3 A person guilty of blackmail shall on conviction on indictment be liable to imprisonment for a term not exceeding fourteen years.

Before the passing of this Act, uttering threats or menaces or writing or causing to be published letters etc which contained threats or menaces against a person, property or reputation with a view to gain or to extort goods or payment or to induce a resisted course of action was conduct punishable quite heavily under various statutes but the crimes were only colloquially referred to as 'Blackmail'. Section 21 Theft Act 1968 collects those crimes under the official designation of 'Blackmail' and provides a broad definition which embraces most of the former law with but minor changes with which security officers should not ordinarily be concerned.

Blackmail differs from robbery in that in robbery the use of threats or menaces relate to action which will put a person in fear of being then and there subjected to force whereas in blackmail the menaces relate to a threat of some future force or to the public disclosure of some fact which would be damaging to the reputation of the victim or the withholding of information with equally damaging effect. It must not be overlooked that the crime of blackmail is committed when unwarranted demands with menaces are made with a view to gain for the person making the demand or for another, or they are made with a view to causing loss to another.

Section 34(2) of Theft Act defines 'gain' or 'loss' to be a material gain or loss in money or other property and goes on to say that 'gain' includes keeping what one has as well as getting what one has not and that a 'loss' includes a loss by not getting what one might get, as well as by parting with what one has.

Security officers should be capable of recognizing an act of blackmail or even an attempt to commit the crime so that something casually heard in conversation, no less than formal complaints, may be properly investigated to root out troublemakers in a business or to bring to light some other unlawful activity which has produced a blackmailing demand. Once recognized it is very evident that investigation should be placed in the hands of police forthwith.

Handling Stolen Goods

Section 22 Theft Act 1968 provides –

1 A person handles stolen goods if (otherwise than in the course of the stealing) knowing or believing them to be stolen goods he dishonestly receives the goods, or dishonestly undertakes or assists in their retention, removal, disposal or realisation by or for the benefit of another person, or if he arranges to do so.

2 A person guilty of handling stolen goods shall on conviction on indictment be liable to imprisonment for a term not exceeding fourteen years.

Section 22 has updated former legislation concerned with the receiving of stolen property and in keeping with other Sections of this Act was simplified and yet extended a definition. The crime now encompasses not only the receiver but all others who in any way handle stolen goods or even provide facilities for stolen goods to be housed or handled between theft and receiver.

By Section 34(2)(b) Theft Act 1968 the term 'goods' 'includes money and every other description of property except land and includes things severed from the land by stealing'. This implies that 'goods' has the same meaning as 'property' which is capable of theft according to Section 4 Theft Act.

By Section 24(1) and (4) Theft Act stolen goods are goods obtained

by theft contrary to Section 1 of the Act

by blackmail contrary to Section 21 of the Act

by deception contrary to Section 15(1) of the Act

abroad in circumstances which would have amounted to theft, blackmail or obtaining by deception if those circumstances had occurred in England.

The goods in question must be stolen goods therefore in most cases of handling stolen goods it will be necessary either for some person to have been found guilty of theft of those goods or for evidence to be forthcoming that the goods were in fact stolen goods and not merely believed to be stolen goods. Goods cease to be stolen goods when they have been restored to the loser or to other lawful possession or custody. (Section 24(3) Theft Act 1968.)

It should be noted that all forms of handling (other than receiving) ie undertaking, assisting or arranging for removal, disposal etc of stolen goods, are subject to the qualification that it must be proved that the offender was acting 'for the benefit of another person'. If he acts for his own benefit he is either a thief, a receiver or concerned with others in the theft of the goods.

In industry and commerce any person who assists theft by transportation, by concealment, or by canvassing the disposal of stolen goods is liable to prosecution for a crime regarded by the law to be more serious than theft and in the same category as burglary.

When management tends to adopt a 'soft' line with persons falling into the above category it would be well for senior security officers to remind management of the seriousness with which the law views such criminal conduct.

Going Equipped to Steal

Section 25 Theft Act provides –

1 A person shall be guilty of an offence if, when not at his place of abode, he has with him any article for use in the course of or in connection with any burglary, theft or cheat.

2 A person guilty of an offence under this section shall on conviction on indictment be liable to imprisonment for a term not exceeding three years.

3 Where a person is charged with an offence under this section, proof that he had with him any article made or adapted for use in committing a burglary, theft or cheat shall be evidence that he had it with him for such use.

4 Any person may arrest without warrant anyone who is, or whom he, with reasonable cause, suspects to be, committing an offence under this section.

5 For purposes of this section an offence under section 12(1) of this Act of taking a conveyance shall be treated as theft, and 'cheat' means an offence under section 15 of this Act.

It should be noted that by Section 25(4) a power of arrest is given parallel with that given by Section 2(2) Criminal Law Act 1967 for arrestable offences. It should also be noted that the offence is committed 'when not at his place of abode' which means that it can be committed, amongst all other places, at his place of work.

This Section is a replacement of the old larceny law relating to the possession by night and without lawful excuse of any key, picklock, crow, jack, bit or other implement of housebreaking. That piece of law was typical of much mid-nineteenth-century criminal law which endeavoured to classify and define specified conduct or items and then found it necessary to generalize so as to encompass anything that the lawmakers had missed from their specification. Modern legislators prefer to use broad terms to embrace all facets of an offence. They do not limit themselves to such expressions as 'an implement of housebreaking' because such an expression gives rise to an argument as to whether or not an article falls into that category – they say 'any article for use in the course of, or in connection with, any burglary, theft or cheat'.

At one stroke the offence is not only extended to thefts and frauds but encompasses all articles of ordinary use where there is evidence to show that it is possessed 'for use in the course of or in connection with' that range of crimes which fall within Sections 1, 9, 12 and 15 Theft Act 1968.

The intention of the old law was to authorise the arrest and prosecution of persons loitering or found to be in suspicious circumstances prior to or preparatory to the commission of housebreaking, shopbreaking etc. That intention has not materially changed and there are circumstances in which it would be right to use the authority of Section 25 to arrest and prosecute when the evidence does not measure up to a theft, a burglary or a cheating but does show that the accused possessed an article with such crime in mind.

An example of the proper use of Section 25 would be if a security guard on duty over a week-end at a large construction site is under instruction to put a stop to a series of thefts of bricks. Following a lengthy period of concealed observation the guard leaves his post to brew fresh tea in his hut and on the way he meets face to face with a man pushing an empty wheelbarrow onto the site. There is no opportunity for observation he is confronting the man and instinctively knows that here is his brick thief but he must have evidence and so he says to the man 'What are you going to do with that barrow?'. Whatever reply is given cannot possibly be a completely satisfactory explanation and so the guard arrests the man for going equipped to steal building material from the site contrary to Section 25(1). The supporting evidence would be the place and direction of travel of the accused with his barrow together with the unlikely and incongruous explanation offered when challenged. It will be for a court to decide whether this ordinary article (a wheelbarrow) was possessed in the course of stealing or whether they find some other acceptable explanation for such possession at that time and place.

There was the case of a woman who was arrested for going equipped to commit burglary whilst in possession of an ordinary dog lead. The facts were that in the London suburb where the arrest took place there had been a series of burglaries effected by someone who managed to open ground floor casement windows of dwelling houses without the use of force. On the night of the arrest two police officers in plain clothes saw the woman with her dog and noticed that she followed the dog into and out of several front gardens of dwelling houses. In their efforts to ascertain her reason for this action they followed her into one garden and found her looking up at the open transom of a ground floor casement type window. The officers asked for an explanation which was not forthcoming and they arrested her leaving the magistrates to decide whether in all the circumstances and having regard to the absence of an explanation they considered that the time and place was sufficient supporting evidence that a dog lead was possessed to open a casement window by hanging it through an open transom. The outcome of the case is not material but that series of burglaries was halted.

'Going equipped to steal', which is the Theft Act title for a Section 25 offence, can embrace a very wide range of articles from a 20 ton lorry driven into a gravel pit to steal gravel to a small scrap of paper purported to be a genuine receipt for payment used to deceive a security guard into allowing goods to pass out of the premises – it includes possession of jemmies and skeleton keys, gloves and adhesive tape, a cheque book and pen and indeed any article when there is supporting evidence to show that possession of such article is in the course of or in connection with any burglary, theft or cheat.

Section 25(3) is important because this sub-section transfers the burden of proof from the prosecution to the defence by stating that 'proof that a person is found in possession of an article made or adapted for use in committing any burglary, theft or cheat shall be evidence that he had it with him for such use'. In effect this means that the prosecution have only to prove possession of such an article and it is for the accused to defend the case by disproving the assumption that he was in possession of the said article to commit crime.

The expression 'article made or adapted' must be taken literally. It means:

a) that the article is one which has been specifically made to commit burglary, theft or cheat – that it is an article for which there is no normal use and that at face value it can only be intended for use in the course of committing burglary, theft or fraud (cheat).

   An example of articles made to commit theft could relate to the possession without lawful reason of metal discs cut to the size and weight of coins of the realm. The discs will probably be used to steal from vending machines or from gaming machines or to defraud the electricity or gas authorities into the supply of power through prepayment meters. Cutting metal discs for those purposes is not uncommon in company workshops and action for a Section 25 offence (proving possession of an article made to commit theft) may be appropriate when there is no immediate evidence that the metal has been stolen on the premises.

   Another example of an article made for use in committing theft would be a copy of a key which key is normally kept under security conditions. An employee, or indeed any person, found to be in possession of such a copy key without a lawful reason could be prosecuted for possessing an article made for use in committing theft and it would be for that person to satisfy the court that there was a lawful and satisfactory reason for such possession. In these circumstances there does not have to be evidence of trespass or clandestine conduct to surround the prosecution's case which will rest upon the presumption that possession of such a key can be for no other purpose than to steal.

   Yet a third example of an article made to commit theft or fraud would be a bank cheque which bears a forged signature. The mere possession of such a document is evidence enough to prove a case under Section 25 without waiting for presentation of the cheque in payment for goods or in exchange for cash, *or*

b) that it is an article of ordinary use which has been altered or revised or changed in form or adjusted so that it now has no use other than to be used in the course of or in connection with any burglary, theft or fraud (cheat). In this context the term adaptation means some physical change and it does not amount to an adaptation of an article merely to use it in the course of committing a crime.

Altering a travelling bag or shopping bag by inserting a false base in which to conceal goods stolen by shoplifting or otherwise unlawfully obtained; or cutting out the bottom of a suitcase so that it can be placed over a smaller case to conceal the smaller case whilst it is stolen are examples of adapting articles to commit theft as also would be the removal of the glass container of a vacuum flask in order to make a larger space for the concealment of stolen goods to be removed from a place of work inside such an innocent looking outer container.

Yet a further example of an article adapted for use in committing theft or fraud would be a normal bank cheque prepared and signed by the account holder but later altered so as to raise the value of the cheque without the knowledge or consent of the account holder. Possession of the document in that condition by any person is evidence enough to prove a charge under Section 25 without evidence of presentation etc and the same may be said of the possession of any document of value so altered or 'adapted'.

Thus to summarize this subject:

1 The Section is useful for dealing with an individual who is caught in possession of any article when there is surrounding evidence to show the intended use of that article in the course of or in connection with any burglary, theft or cheat and prior to or whilst he/she is preparing to commit such a crime.

2 The offence may be charged if the article possessed is one specially made to commit a burglary, theft or cheat as illustrated above.

3 The offence may be charged if the article possessed is an ordinary article which has been physically adapted for use in committing a burglary, theft or cheat and which in that condition has no ordinary use.

In the case of either 2 or 3 above it is necessary to invite an explanation for the person's possession of the article before proceding to a prosecution because the absence of a lawful reason creates the presumption that the article is possessed for the purpose of crime.

One last point to particularly note is that Section 25(3) is an administrative direction which does not create a fresh offence and that there is only one charge which is contrary to Section 25(1). The wording of a formal charge under Section 25(1) should include a description of the article in respect of which the charge is made. In appropriate cases the evidence given in support of the charge could infer that the article had been made or adapted for burglary, theft or cheat and that it had no ordinary use but guidance on this point would be necessary from a legal representative or the police.

Other Crimes and Provisions of Theft Act 1968

Limitation of space does not permit for further comment here upon other crimes and provisions contained in this very important Act and readers are advised to study the Act closely in order to acquire a proficient krowledge of the law of theft and related crimes.

CHAPTER 17

Other Offences and Trespass

Assault and Breach of the Peace

It is most important that security officers should know and understand where they stand in law when personal violence is used upon others or upon themselves.

An assault is an attempt or offer by force to do bodily hurt to another. It is not necessary that the other person should be touched and if the offer of violence is in circumstances in which the assailant has the means and ability to carry out his threatened intention then an assault has been committed. A battery is the actual application of that force upon the other without lawful excuse. Every battery follows an assault since the assault becomes a battery when the blow is struck. It is common practice to speak of an assault when in fact one is speaking of an assault and battery.

Raising the fists or a cudgel at another is an assault, throwing stones or other injurious articles at another is an assault whether or not the stones etc strike the other. Mere threatening words do not in themselves constitute an assault but if several persons gather round another offering by word or gesture to do him some hurt so that he is in fear of physical violence then each such member of the crowd is committing an assault upon him. Setting a dog to attack another may amount to an assault if the dog causes physical injury to the other, hence the Home Office warning that 'no dog should be used for security purposes unless it is fully and properly trained to such a standard that it can be kept under adequate control at all times'. (Home Office Code of Practice for the Use of Guard Dogs 1975. See also Chapter 22.)

An assault may be committed during 'horseplay' such as when a forklift truck is driven straight at another who in fear jumps out of its path or when a bucket of water is tilted over a doorway through which the 'victim' is known to be passing or by pulling away a chair or seat upon which someone is about to sit so that he falls to the ground. Insulting, provocative and abusive words or conduct do not amount to an assault nor are they justification for an assault. Prompt action should be taken to stop the use of such words or conduct in working areas and the aggressor should be warned that his conduct might amount to a breach of company rules and working conditions. Such matters should be promptly reported to heads of departments who are responsible to the company for the conduct of their employees.

Assaults are said to be justified ie excused by the Law, when they are committed –

1 in self defence or defence of one's own property. It is not necessary to confine oneself to warding off blows but one may hit back if that be necessary to defend oneself from harm provided that the amount of force used is no greater than that offered by the assailant. Defence must remain defensive and not become an attack. Restraint is the best policy,

2 under the authority of law as for example to physically detain, or to restrain from escape, a person who is being arrested under the authority of Section 2 Criminal Law Act 1967 or any other legislation giving authority for any person to arrest another without a warrant. The Children and Young Persons Act 1933 gives authority to a parent or teacher to administer reasonable and moderate chastisement of a child;

3 during a lawful game or sport when physical contact between participants is made within the rules of the game or sport e.g. boxing, wrestling, rugby football, fencing etc. Prizefighting is a series of assaults because there are no rules therefore is a breach of the peace and unlawful;

4 by misadventure or accident but note the comments above about horseplay in which the 'injury' sustained or intended is deliberate and therefore is not accidental.

The law classifies the seriousness of assaults according to the amount of violence used or intended by the assailant at the time of committing an assault or by the extent or severity of the injury sustained by the victim so that a push in the chest resulting in a fall and a fractured skull may be charged as causing grievous bodily harm whilst an attack with a cudgel which is thwarted by the intervention of a third party may result in a conviction for aggravated assault only.

The general division of assaults is as follows:

i *Common assault* Section 42 Offences against the Person Act 1861

In minor skirmishes where each party takes out a summons against the other under Section 42 the courts will usually find both cases proven and will bind over both parties to be of good behaviour for a given period. This allows time for tempers to cool. If an injured party complains to the court that any of the conditions of a bindover are being breached the defaulter may be summoned before the court to show cause why the amount of his bindover should not be estreated (paid into court). If the court orders estreatment this may be enforced by commitment to prison for non payment.

ii *Assault occasioning actual bodily harm* Section 47 Offences Against the Person Act 1861

It has been held that such an assault is one in which no wound or other serious injury is inflicted but which interferes with the health or comfort of the victim.

iii *Unlawfully and maliciously wounding* Section 20 Offences Against the Person Act 1861

A wounding is a battery by which the flesh is opened. A mere scratch may constitute a wounding.

iv *Unlawfully and maliciously causing any grievous bodily harm with or without a weapon or instrument* Section 20 Offences Against the Person Act 1861

An injury falls into the category of grievous bodily harm if there is a serious interference with the health or comfort of the victim.

v *Unlawfully and maliciously causing to any person any grievous bodily harm with intent to do some grievous bodily harm to any person* Section 18 Offences Against the Person Act 1861

This is the serious one when not only is gbh caused to someone but when there is evidence that the kind of attack or the weapon used demonstrated the assailant's intention to cause such an injury to that person or to any other person. The maximum penalty under Section 18 is imprisonment for life.

vi *Assault with intent to commit an arrestable offence; assault upon a police officer in the execution of his duty or upon any person acting in aid of such an officer; or assault upon any person with intent to resist the arrest of himself or any other person* Section 38 Offences Against the Person Act 1861

Careful note should be taken of this protection given to civilians acting under the authority of law. The Section also makes it an offence to commit any kind of assault upon those assisting police or endeavouring to maintain law and order.

vii *Unlawfully and maliciously causing gbh with intent to resist or prevent the lawful apprehension of any person* Section 18 Offences Against the Person Act 1861

viii *Assaults involving indecency* on males and/or females by males and/or females formerly were covered by various sections of the Offences Against the Persons Act 1861 but more latterly by the Sexual Offences Act 1956.

All indecent assaults except rape by a man upon a woman are considered to be lesser crimes and, apart from prompt verbal reporting of a complaint made to them, no other action should be taken by civilian security staff in such cases.

ix *Assaults upon children and young persons*

Charges framed under appropriate Sections of the Offences Against the Person Act 1861 are allied to . . .

The Children and Young Persons Acts which are concerned with stepping up penalties where the victim is a child or young person and with the admissibility of evidence which in other circumstances may be hearsay.

x *Homicide* Section 1 Homicide Act 1957

There is much common law and case law defining justifiable and excusable homicide but these are matters for the police investigators and the Courts. To security personnel a dead body in any circumstances means calling police as quickly as possible, preserving the scene, ensuring that witnesses do not depart, and informing management by the most expeditious means.

It should be noted that all crimes falling to be dealt with under the following Sections are arrestable offences for the purpose of Section 2 Criminal Law Act 1967:

Offences Against the Person Act 1861 Sections 18, 20, 38 and 47.
Rape at Common Law.

In law security personnel have only the same powers and authority as those derived from common law by all citizens whose responsibility and duty of citizenship it is to keep the Queen's Peace, to obey the law and to ensure that others do so. If an assault is seen to be taking place it is any security man's duty as a citizen to intervene to stop the assault or continuation of the assault and to prevent a counter-attack from the person first assaulted. He may use such force as may be necessary to sustain his intervention without personally becoming liable in law for the force he is using.

If employees complain to security that they have been assaulted they should be referred to the police who will advise them as to their remedy in law or take up an investigation, and maybe a prosecution, on their behalf according to the nature and severity of the assault. If the assault has been of a serious nature police should be called to the premises on behalf of the complainant. If a knife, bludgeon or other weapon has been used the assailant should be given no opportunity of escape before the arrival of police who should be called by the security staff irrespective of whether the victim asks for them to come or not. If the victim is unconscious the police must be called in addition to medical aid and an ambulance. In these serious cases nothing in the surroundings where the assault took place may be moved otherwise than to remove the victim from risk of further injury or to remove from the assailant his ability to continue the assault or to make good his escape. Any person in the vicinity whether a witness of the assault or not, or any other person who can provide information of what led up to the attack should be asked to remain in the vicinity until the arrival of police. Senior management must be notified forthwith. Full use must be made of notebooks to record times, identification details of persons involved and witnesses, the exact position of any property connected with the incident, and any remarks or statements about the incident made in the presence and hearing of the assailant as well as his replies or responses to such statements. All such information should be given to the police when they arrive. The reader's attention is also called to Chapter 18 on the subject of identification of property at the scene of a crime.

Assaults committed at work may be regarded as wilful misconduct during employment and could result in dismissal regardless of any action taken in the Courts by or on behalf of the person assaulted. Every case must be carefully investigated within the company to ascertain whether there has been provocation or even justification for what has taken place. A claim of self-defence might be a very good answer to a complaint of assault particularly if it is supported by witnesses.

The wrongful action of committing an assault and battery upon another is one of those 'torts' which may be dealt with by civil law or by criminal law or by both. If the assault is one considered to be worthy of punishment then prosecution of the assailant will be taken in the criminal courts by the police or by the victim. Upon a conviction the court of trial may, in addition to passing sentence as a punishment upon the accused, order him to pay compensation to the victim within sums fixed by the law. Should the court of trial not make an order for compensation the legal advisers for the victim may advise him or her to sue the assailant in the civil courts for compensation and the sums awarded in these courts may be very much higher than those permitted in the criminal courts.

Civil court proceedings for compensation may also be taken against an assailant by or on behalf of the victim despite the fact that no criminal prosecution has been taken. Should it be decided at the outset that both criminal and civil proceedings are contemplated the criminal proceedings must be completed first.

When injuries have been received as the result of some criminal action on the part of another or whilst assisting police to prevent a crime or to arrest a criminal, compensation may be received from the Criminal Injuries Compensation Board of 10/12 Russell Square, London WC1.

Breach of the Peace and Threatening Behaviour

A 'breach of the peace' takes place when the normal peace of society is interrupted as for example by an assault, or by an affray (ie two or more persons engaged in a fight to the terror of the Queen's subjects), or by a serious incident involving crowds such as in an unlawful assembly of persons or in a riot. Courts may bind over to keep the peace persons who have been involved in, or who are known to have intentions of committing, a breach of the peace or of organizing an activity which is likely to cause a breach of the peace and failure to comply with the court order renders the offender liable to imprisonment for contempt.

Persons who use threatening, abusive or insulting words or behaviour in any public place or at a public meeting render themselves liable to prosecution. The relevant statute is the Public Order Act 1936 Section 5 as amended by Public Order Act 1963 which provides:

> Any person who, in any public place or at any public meeting, (a) uses threatening, abusive or insulting words or behaviour, with intent to provoke a breach of the peace or whereby a breach of the peace is *likely* to be occasioned; or (b) distributes or displays any writing, sign or visible representation which is threatening, abusive or insulting, is guilty of an offence punishable summarily with imprisonment for three months or a fine of £100 or both, and, on indictment, with imprisonment for twelve months or a fine of £500 or both.

Generally speaking the Public Order Act is only used to deal with serious or political threats to the peace and local statutes and byelaws containing somewhat similar provisions are usually invoked to prosecute those who are disorderly and aggressive towards others in the streets and public places. An example is to be found in the Metropolitan Police Act 1839 Section 54 (as amended) which reads:

> Every person who in any thoroughfare or public place shall use any threatening, abusive or insulting words or behaviour with intent to provoke a breach of the peace or whereby a breach of the peace *may* be occasioned shall be guilty of an offence.

The maximum fine on summary conviction is £10.

Power of arrest is given *to police* to arrest offenders under either of the above statutes and similar powers apply under local laws in other cities and towns. The expression 'public place' has been defined by Section 33 Criminal Justice Act 1972 as, 'any highway and any other premises or place to which at the material time the public have or are permitted to have access whether on payment or otherwise'. Ordinarily one does not regard factory, commercial or other business premises to be a public place but circumstances may change so that a sports ground which is open to the public for a fete, or a business area which is thrown open for inspection by the public as an 'open day', could become a public place for the duration of the event. Football grounds, dog racing arenas and halls used for boxing events fall naturally into the definition.

The offences are complete if words or behaviour are used which are either threatening or abusive or insulting. It is usually a practical deterrent for offenders to be warned that continuance of their conduct could involve them in a prosecution which may be taken against them by way of a summons whether or not police are called to the scene.

When dealing with these cases a court may, in addition to imposing penalties, bind over the offenders to keep the peace for a period named in the bindover.

Prompt intervention should be taken by Security Officers to subdue threatening, abusive or insulting language or conduct and to prevent a more serious disturbance. Particulars of aggressive parties should always be reported to management.

Criminal Damage

The Criminal Damage Act 1971 is a very short Act the major parts of which are set out below. In a few short paragraphs this Act brings up to date and reduces a mass of former law dealing with wilful and malicious damage whilst at the same time extending the law to take into account the contemporary point of view with regard to one's rights over one's own and other people's property.

The serious results which can come from damaging any property by fire has caused a change in the law relating to arson. Whereas the old common law crime of arson was related only to setting fire to dwellings arson is now defined as causing any criminal damage as defined in Section 1(1) and (2) of the new Act by means of fire. All arson is punishable by imprisonment for life.

By Section 1(1) Criminal Damage Act 1971 a person who without lawful excuse destroys or damages any property belonging to another intending to destroy or damage any such property or being reckless as to whether any such property would be destroyed or damaged shall be guilty of an offence punishable by imprisonment for a term not exceeding ten years.

By Section 1(2) Criminal Damage Act 1971 a person who without lawful excuse destroys or damages any property, whether belonging to himself or another

a) intending to destroy or damage any property or being reckless as to whether any property would be destroyed or damaged; and
b) intending by the destruction or damage to endanger the life of another or being reckless as to whether the life of another would be thereby endangered;

shall be guilty of an offence punishable by imprisonment for life.

These two sub-sections may be summarized as follows:

Section 1(1) creates the offence of damaging or destroying *another's* property with intent to damage or destroy that property or being reckless in that regard.

Section 1(2) creates the aggravated offence of damaging or destroying *any* property in circumstances as before AND intending by such action to endanger the life of another or being reckless in that regard.

Note the constituent parts of each sub-section of Section 1. It is no crime to damage one's own property wilfully or recklessly unless the intention of such damage is to endanger the life of another person or the circumstances in which the damage is committed are so reckless as to be regardless of danger to the lives of others. For example, destroying with explosives one's own motor car in a busy shopping area would be a crime by Section 1(2) whether or not persons were killed or injured thereby but to destroy the car in that way in an isolated spot would be no crime against this Act.

By Section 2 Criminal Damage Act 1971 a person who without lawful excuse makes to another a threat, intending that the other would fear it would be carried out

a) to destroy or damage any property belonging to that other or a third person; *or*
b) to destroy or damage his own property in a way which he knows is likely to endanger the life of that other or a third person

shall be guilty of an offence punishable by imprisonment for a term not exceeding ten years.

This Section is an extension upon former law and amongst other matters makes a person criminally liable for his actions when making threats over the telephone to cause damage by placing of a bomb so that the person receiving the telephone call is fearful that the threat will be carried out regardless of whose property may be damaged by the bomb. It is immaterial whether the person making the threat intends to carry it out. The Section also establishes as a crime the action of a disgruntled employee who, during an argument with his manager or foreman, threatens any violence towards plant or product in such a way that the manager or foreman fear that the plant etc will be damaged or destroyed. The penalty under the Section makes such action an arrestable offence so that immediate action can be taken to prevent the employee from carrying out his threat.

By Section 3 Criminal Damage Act 1971 a person who has anything in his custody or under his control intending without lawful excuse to use it or cause or permit another to use it

a) to destroy or damage any property belonging to some other person; or
b) to destroy or damage his own or the user's property in a way which he knows is likely to endanger the life of some other person;

shall be guilty of an offence punishable by imprisonment for a term not exceeding ten years.

By this Section the employee making threats to destroy plant or product commits a further offence if his threats are accompanied by the flourish of a heavy spanner, crowbar or like weapon of destruction. These illustrations are of course figurative but can be likened to an abundance of situations.

Many people commit small acts of criminal damage without realizing the gravity placed upon their actions by the law. For instance it is not uncommon for workpeople to destroy by fire or to smash an article instead of cleaning and re-using it – 'There are plenty of new ones in the stores' is their outlook and they would be shocked or even outraged to be told that what they have done is criminal and punishable by imprisonment. It is so easy to be reckless with other people's property without a 'feeling' of responsibility. Maybe they should be reminded now and again of their responsibility and that is a task for security.

Section 5 Criminal Damage Act 1971 qualifies the meaning of the words 'without lawful excuse' as used in Sections 1(1), 2(a) and 3(a) of the Act and broadly provides a statutory defence to a charge under those particular Sections if the court of trial is satisfied (a) that the accused held an honest belief when committing the offence that the person entitled to consent to the destruction or damage had given consent or would have consented to the damage or destruction if he had known of it and its circumstances; or (b) held an honest belief that he was protecting property belonging to himself or another whether that belief was well founded or not and provided he believed that the property was in immediate need of protection and that the means he was adopting were reasonable in the circumstances.

It should be noted that Section 5 does not apply to those clauses in Sections 1(2), 2(b) and 3(b) which relate to the causing of damage etc or possession of an article for that purpose when there is an intent to endanger life or when there is recklessness in that regard.

It would appear that a lawful excuse would exist when a person causes damage by breaking into premises to quench a fire which he knows has started inside. Against any subsequent complaint of unlawful damage and interference with the premises that person could claim an honest belief that the owner of the premises would have consented had he known of the fire. There would also appear to be a lawful excuse for damage caused to a building to rid it of woodworm, dry rot or vermin even though the owner is abroad and out of contact for a year – such damage being caused in the honest belief by the person responsible for causing the damage that the methods he employed were necessary to save the rest of the building from further harm.

It is of immense interest to students of the criminal law that in Section 5 of this Act as in Section 2 of the Theft Act 1968 the lawmakers have attempted to keep out of the Courts those prosecutions which in common-sense terms should not be taken if the investigators look carefully enough into the causes for the commission of the alleged crimes.

An aggrieved party may always seek redress and compensation from a perpetrator of damage in the Civil Courts without criminal proceedings having been taken but if criminal proceedings are to be taken they should be taken before any Civil Court action.

Section 6 Criminal Damage Act 1971 authorizes the issue by magistrates of warrants to search for any article used for or intended to be used to cause damage etc. An application for such a warrant would normally only be made by the Police.

Section 8 Criminal Damage Act 1971 provides that compensation for the cost of making good damage or destruction may be ordered by a Court before which a person has been convicted of any offence under Section 1(1) or 1(2) of the Act. The maximum order may be for £400 in a Magistrates Court but there is no limit in higher Courts.

Section 10 Criminal Damage Act 1971 defines 'property' for the purpose of the Act as property of a tangible nature, whether real or personal, including money and –

a) including wild creatures which have been tamed or are ordinarily kept in captivity, and any other wild creatures or their carcases if, but only if, they have been reduced into possession which has not been lost or abandoned or are in the course of being reduced into possession; but
b) not including mushrooms growing wild on any land or flowers, fruit or foliage of a plant growing wild on any land.

For the purpose of this sub-section 'mushrooms' includes any fungus and 'plant' includes any shrub or tree, and by sub-section (2) goes on to say property shall be treated for the purposes of this Act as belonging to any person –

a) having the custody of it;
b) having in it any proprietary right or interest (not being an equitable interest arising only from an agreement to transfer or grant an interest; or
c) having a charge on it.

Where property is subject to a trust, the person to whom it belongs shall be so treated as including any person having a right to enforce the trust.

Property of a corporation sole shall be so treated as belonging to the corporation notwithstanding a vacancy in the corporation.

The two definitions in Section 10 bear very close relationship to similar definitions in Theft Act 1968 Sections 4 and 5 and comparison will assist students to clarify each such definition in his own mind.

The use of the term 'reckless' in this Act is of interest and is already defined in law as the state of realising that harmful consequences of one's action can occur but being indifferent as to whether it does or not and this includes deliberately taking a foreseen risk.

Comment upon the Criminal Damage Act would not be complete without drawing the reader's attention to the fact that whilst the Act itself does not provide a power of arrest penalties under the Act are such that all offences under the Act are 'arrestable offences' for the purpose of Sections 2 and 3 of Criminal Law Act 1967. The civilian's power of arrest for arrestable offences is discussed in Chapter 19.

In the field of commercial and industrial security the Criminal Damage Act may not be frequently invoked but it will be of extreme value for dealing with acts of vandalism committed for reasons of sabotage or spite if the current trend towards violence and the rejection of authority proliferates.

Trespass

In Chapter 14 distinction was drawn between the civil law and criminal law and it was said that whilst most wrongs fall quite clearly to be dealt with in the civil courts or the criminal courts there are a few wrongs which may fall to be dealt with alternatively by civil law or criminal law and sometimes by both. We have already discussed a similar proposition under assaults and whilst it may be said that trespass, like assault, is both a civil and criminal tort the law concerning trespass is somewhat more complex.

Trespass is an entry upon the land or premises of another without consent or without a right or authority to be there, or remaining upon the land or premises of another when an original consent has been withdrawn or no longer exists. Trespass may also be committed by a person authorised to be in one part of a building moving into another part where he has no authority to be.

If consent to entry has been obtained by some trick or fraudulent story the entry is a trespass as for example if entry has been obtained by a bogus meter reader – bogus telephone engineer – bogus market researcher or bogus policeman. By the authority of Section 2(6) Criminal Law Act and for the purpose of making an arrest of a person for an arrestable offence, *a constable* may enter, if need be by force, any place where that person is or where the constable suspects him to be. Any other entry by force is a trespass unless it is authorized by court warrant. A trespass may be committed by the entry of a vehicle onto private land, or by the planting of a tree, or by the erection of a fence, or by dumping rubbish upon the land of another. Under special circumstances the mere insertion of a finger to release a latch or of a piece of wire or hook may be regarded as an entry and a trespass if without consent.

Trespass may be the innocent action of someone who unwittingly is on private premises without a right to be there or it may be the wilful action of someone who enters upon or remains upon private land or premises as an encroachment upon the rights and privacy of the owner of that land.

Mere trespass is not a criminal wrong – is not punishable and trespassers cannot be prosecuted just for the trespass but a trespasser must remove himself or the means by which he is trespassing at the first request of the owner of the land or premises or upon a request to leave made on behalf of the owner by anyone acting as his servant or agent. If the trespasser does not peaceably leave he may be removed from the premises by the owner or by his servant or agent who is authorized by the common law of the land to use such assistance and such force as is necessary to effect that removal. Thus if caravan dwellers enter a field to park their caravan they may be asked to leave by or on behalf of the owner and if they do not do so the owner or his servant or agent may by themselves or by employing some other person tow the caravan from the field to the nearest public highway; if a fence is erected upon what is supposed to be the boundary of land but which is within the property of one of the landowners that landowner may remove the fence provided he does no more damage than that necessary for the removal and does not retain any of the fencing or posts after its removal; if squatters enter an unoccupied house to live there the owner may turn them out if he can do so in a peaceable way but reference will be made to this situation again later in the chapter.

The owner of land or premises aggrieved by the action of a person who is committing the tort or wrongful act of trespass may, as an alternative to the ejection of the trespasser, apply to a civil court for an order to be made directing the trespasser to remove himself, or that by which he has entered the land or premises. If the trespasser refuses to comply with the order the court may direct other persons to remove the trespass and may commit him to prison for his contempt of court.

Taking the three examples listed above the caravan dwellers who refuse to leave the field may be ordered by a court to leave and if they do not do so will have their caravan towed out by a contractor working under the authority of the civil court order; the person responsible for erecting the fence upon the land of another may be ordered by a court to remove it and if he does not do so the court may employ persons to remove it; and the squatters who refuse to leave the empty house may be ejected by an officer of the court and he may call upon police to accompany him to preserve the peace and to protect him from assault whilst executing the order of the court.

Notices are sometimes erected on land or premises warning persons not to trespass and these are quite in order but should not include a threat of prosecution. Notices should be informative as for example:

'Private Property – Keep Out'
'Do not trespass on this land'
'Trespassing prohibited by Blanco & Co'
'No entry without special approval. Enquire at the Gatehouse'
'Ticketholders only'
'If you pass this sign, you will be trespassing'
'Persons loitering on this staircase will be dealt with as trespassers'

So many people have the wrong idea about their entitlement to be on other people's premises and wrongfully assume that if there is not a notice barring their entry that entry is lawful. Restauranteurs, shop-keepers and publicans who openly invite the public to enter may yet bar access to their premises to individuals without assigning a reason and of course admittance to clubs and such like places is limited to those holding membership and their guests. People at work only have a right of access to their places of work for the purposes of their employment and will be trespassing if they enter a place of work without specific approval at any other time or for any other purpose. When employees withdraw their labour during an industrial dispute they automatically cancel their right to be on the premises and may re-enter only upon settlement of their grievance or otherwise by consent of the management.

Whether notices are displayed or not any person who is on company premises without lawful business there or without permission of the company, or any of its employees, is a trespasser and should be challenged as to his reasons for being there. He may have entered for a perfectly legitimate purpose but then refuse to leave when that purpose has expired, for example, he may have a dispute with the company and decline to leave peacefully after an abortive interview. It may be that entry has been gained by some subterfuge such as a purported delivery to the main stores or the canteen but in fact the intruder is there to sell razor blades to employees in the canteen during the meal break time. Entry may have been obtained by falsely claiming to have an appointment with someone on the staff or the intruder may just have walked in from the street because there was no challenge at the threshold and is in fact there to photograph the plant, or the layout of the premises for some ulterior motive. In these days of industrial espionage, planted bombs, cash office raids, dubious sub-sub-contract workers, thefts of all kinds including the theft of laden vehicles from inside of works premises, and of numerous other criminal and malicious activities the presence of any stranger within the perimeter should be regarded with suspicion and he must be questioned.

When making a challenge of a stranger the security officer's first action should be to invite the stranger to accompany him to the security office where the officer will be in favourable surroundings of colleagues, telephones, good light and the means to keep the stranger from departing if necessary. Questioning a person on a staircase or landing or in a darkened yard is not good security action. The simple action of asking him to go to the office could result in the stranger taking to his heels to run out of the premises whereupon he should be caught, restrained and escorted to the office.

In the office his identity should be asked for together with support for the information given such as the production of driving licence, club cards, credit cards, pension book or similar items carried by normal well-intentioned persons. An inability to produce anything of that nature suggests that he has come prepared to supply no verification of an untruthful name and address. He should next be asked to give an explanation of why he was in the place where stopped and why he was moving in the direction in which he was seen to be going. Careful consideration must be given to that explanation and it must not be accepted as genuine on its face value or even to be truthful unless all surrounding facts known to the officer support it as a genuine explanation.

It is only after a security officer has satisfied himself beyond reasonable doubt that there can be no unlawful intent in an intruder's presence upon the premises that he may regard him as a mere trespasser and ask him to leave.

With the passing of the Theft Act 1968 the word 'trespasser' became significant because of its use in Section 9 in the revised definition of Burglary. The maximum penalty on indictment for burglary is imprisonment for not more than 14 years and he would be a sorry security officer who led off the premises as a trespasser a would-be burglar of whom he had not asked sufficient questions about his presence there when in fact the officer had actually caught him 'on the job' by 'having entered the premises as a trespasser with the intention of committing theft or unlawful damage etc'. It is the evidence of the intruder's intention that is important and which quite often can be obtained by astute questioning or by examination of property in his possession. Such property might include a key to fit the ignition of one of the laden lorries in the yard or a pair of wire cutters with which to enter the valuable goods compound! Readers may ask 'What power is there to search such a person?' and the answer is 'None without his consent' but asking him to turn out his pockets is not searching him and if he declines to oblige, there are grounds for calling the police to assist.

Factors which should influence a security officer's judgement as to whether the man is a mere trespasser or someone more sinister should be the time of day or night and the part of the premises where the intruder was found – information in possession of the officer as to the man's movements or any conversation he has had with other persons before challenged by the officer – his frankness in answering questions or his willingness to co-operate and to turn out his pockets or to allow an inspection of a vehicle he has in the vicinity. Suspicion should fall heavily upon the intruder who is most apologetic from the start and who is only too willing and anxious to leave even before this is suggested to him – such a person must not be allowed to leave without an investigation and it is for him then to continue his policy of co-operation to the officer's entire satisfaction or, if he changes his attitude under a little pressure, once again there are grounds for calling the police.

The citizen's right to make an arrest is dealt with in Chapter 19 and until the reader clearly understands that chapter it is not easy to enter into a discussion here about arresting trespassers who do not give satisfactory explanations for their presence on company premises but about whom there is no evidence of suspicious behaviour and who have no articles in their possession which could give grounds for accusing them of criminal intent.

A mere 'hunch' or belief that he is on the premises for a malicious motive is not good enough to permit him to be arrested unless it is backed up by other evidence to indicate the unlawful purpose of his intrusion such as:

his general conduct before and during interrogation; *or*
his evasive answers to questions; *or*
his remarks to others before security intervened; *or*
some article which is in his possession.

If, and only if, evidence of a tangible nature is available or forthcoming during questioning and investigation to show an unlawful purpose for his presence on the premises may the authority of Section 4 Vagrancy Act 1824 be used to arrest him for being found on enclosed premises for an unlawful purpose (and that usually means to steal.) Without such evidence he must be allowed to leave. Alternatively, delaying tactics should be used such as a prolonged verification of the explanation he has given, or the inability to get through on a telephone call to clarify his story, such delay being used to give the police time to answer a call for assistance. Thereafter the enquiry passes into the hands of the police and the security officer has not become involved in declaring an intention to arrest him which may or may not be a lawful action and which may or may not have involved the use of force to hold him against his wishes.

The common law right for an occupier or owner of premises to eject a trespasser by force if necessary has taken a new slant in recent times because of actions taken in the courts against owners and their agents who have employed strong arm methods to regain possession of premises occupied by squatters. It cannot be too strongly emphasised that squatters may not be removed from dwellings without an order of a court and that harassment and eviction even of unlawful squatters are serious offences.

This does not mean that they are immune to the ordinary laws of theft and criminal damage. Employees who occupy premises or buildings and refuse access by normal staff and management in order to enforce a settlement of their demands relating to pay and conditions are trespassers who anticipate that management will be unlikely to exacerbate the situation by using force to evict them. The lesson to be learned is that no opportunity should be given for squatters to enter and take over a building and in this respect a serious resonsibility rests upon security to be alert for such a move and to promptly lock up vacated premises at the close of normal business.

The intruder who enters marginal land to set snares or to shoot birds commits specific offences under the Game Laws or maybe under the Firearms Act and in connection with such laws the police appreciate the display of proper notices indicating that the land is privately owned and that trespassing there is forbidden by the owners.

Where members of the public form the habit of crossing part of company premises as a means of a short cut access to adjoining premises, action must be taken to stop them because such habitual use could create a right of way. Should barriers be impracticable, notices should be displayed at both ends of the 'route' indicating that any person crossing the land is trespassing unless prior consent of the company or of one of its authorized employees has been obtained for each and every such crossing. Thereafter an occasional stopping or refusing of access would diminish a claim of right by custom and habit.

Trespass is a summary offence (that means an offence punishable by law in a Magistrates Court) when committed upon land protected by Act of Parliament or by Bye Laws or by Statutory Instrument in a similar manner to the protection given to railway premises. Such protected land may be mines, docks, airfields, river and canalways, HM Forces establishments and similar Government property. In the absence of such protection and, if the trespass is not committed with the intents mentioned in Section 9 of the Theft Act or if there is no *evidence* of his unlawful purpose under the Vagrancy Act, mere trespass is not a punishable wrong.

CHAPTER 18

The Investigation of Crime and Other 'Incidents'

By asking the right or wrong questions
of the right or wrong people
in the right or wrong way
a security officer may make or mar a good investigation from the very outset of enquiries. He may also place his employers in a sound legal position or in a position of profound embarrassment according to whether he has or has not acted with strict impartiality, sound reasoning, and correct decorum. Those mighty phrases aptly sum up the responsibility placed upon security officers in their role of investigators and they will not receive much shelter from criticism nor consolation from management or colleagues if they make an unholy hash of an enquiry.

The ways by which an enquiry may be efficiently handled thereby avoiding criticism are dealt with in some detail in this chapter but putting good advice into practice requires special qualities in the investigator. He must have good powers of observation, good commonsense, patience and perseverence, he must be methodical in thought and action and possess a broad outlook upon life and its human problems to which he applies a logical approach. He must have a sound knowledge of law and practice affecting his sphere of operations and should understand the demands and rules of evidence. He must be capable of acting independently with confidence and determination whilst prepared at the same time to accept co-operation and guidance from those qualified to give it.

Any instruction which is given on the subject of investigation must of course be regulated by the wishes expressed in standing orders or by the written or known policies of each individual company. This will vary according to the company's attitude towards the involvement of civilian police in company affairs; in their attitude towards company and employee relations; and in the depth of responsibility which they give to members of their security department. Therefore the views of the authors expressed in this chapter may be accepted or challenged according to those company wishes.

Whilst it is assumed that the majority of investigations at work will be carried out by senior security people, all ranks of a security department should be prepared to act on their own initiative at least to deal with the initial stages of an enquiry.

If serious crime has taken place the calling of police should be a standing order and there should be no delay occasioned by an internal investigation otherwise than to establish reasonable cause to believe that such a crime has been committed or is genuinely alleged. The emergency procedure for calling police may be used if necessary. Serious crimes include fraud, forgery, falsification of accounts, corrupt practice, bribery or blackmail; serious assault, wounding, robbery, serious theft or burglary; criminal damage amounting to sabotage or arson or any offer of violence to persons or property by the use of explosives, firearms or other weapons of attack.

Firm steps should be taken to protect and preserve all evidence and to protect a scene of a crime from busybodies and autocratic middle managers who feel the need to get themselves involved. If there are allegations of such serious crime made against individuals those individuals should be prevented from making good an escape or of leaving under any pretext or excuse before police arrive. Complainants, informants and witnesses should be asked not to leave the premises pending arrival of police. They may and should be kept apart from any suspect or accused person and each should be identified to the police upon arrival.

Police should, if possible, be met by the security officer in possession of the facts in order to minimize delay and to avoid confusion from the outset. The officer should clearly inform the police what he knows of the facts and what has been done so far and from that moment on the investigation will be in the hands of the police who should be given full co-operation and access to places and matters which they regard to be of value in the furtherance of their enquiries. There is no point in calling police if they are to be refused access to vulnerable places or to documentation or information. It is unreasonable to call them and then to place restrictions upon their actions, to decline to give them facilities for carrying out their business or to fail to back up any prosecution they wish to take.

If it is company policy that the security officer will carry out an investigation in every case before police are called the officer will doubtless be a senior person with much experience in the work and no advice here given is likely to advance his knowledge or technique but if a company does not employ such an experienced officer then in those cases already listed as serious crimes it would not only be unwise but probably harmful to police enquiries for any but the most preliminary investigation to take place before police are called. In all other cases internal investigation is permissible and in most cases will be essential in order –

1 to establish that the offence alleged has in fact been committed;
2 to establish the identity of the offender;
3 to obtain evidence upon which a prosecution may be taken or upon

which company disciplinary action may be taken in addition to, or as an alternative to, prosecution.

The security officer's investigation should take the following pattern –

1 He must keep in mind that he is a private person with no special powers or authority to question anyone, to require answers or to direct things to be done otherwise than within the orbit of the authority entrusted to him by his employer and described in his standing orders.

2 His first duty is to *establish the facts* – what has happened – who has been involved so far – how much of what is being said or alleged can be proved to be correct and how much is fantasy, imagination, belief or sheer guesswork. He must sift the facts from the fiction and decide what further facts need to be proved to establish a wrongdoing and a wrongdoer. Where exactly was the property last known to be or last seen? From where exactly was the property taken? When was that? Precisely what has been taken or lost or damaged and how can it be identified? Is it now available for inspection and identification? How has the loss become known and over what period or at what precise time did it occur? How is the accused known to have contact with the article or have had the opportunity for contact with the article – who is alleging this and what substantiation or verification is there of that allegation?

3 He must clearly identify the informant and/or complainant and fully question him to ensure that his story is genuine and not malicious – is supported by circumstances which are correct – is not borne of mistake or prejudice. If the complaint is about an assault, does the complainant (the victim) understand that an investigation and any outcome of that investigation by way of criminal, civil or disciplinary proceedings taken against the assailant will be on behalf of the complainant and of him alone.

In a one against one situation the investigator must not just drop the enquiry because he does not know which side to believe, he must conduct and pursue his enquiry of the complainant, informant, loser, accuser, whichever he is dealing with to establish in his own mind that there is some justification for the complaint; or that the information is from a reliable person who has no axe to grind; or that the loser really was in possession of the missing article in circumstances which could have given an opportunity for the accused to steal it; or that an accuser is not acting out of malice but has a genuine belief in what he asserts; or that there is corroboration from some other circumstance to support what has been alleged.

4 He must consider the advisability of forthwith taking down in writing as a statement what is said by complainants, informants and witnesses whilst their story is still fresh in their minds and before it becomes distorted by what they may hear from others. Once written down the statement is always available for reference and for cross checking as the investigation proceeds.

5 He should widen the scope of enquiries to identify witnesses and to seek more information from witnesses or others who may know some facts which they do not consider to be important enough to mention, to bring out knowledge which reluctant persons are not willing to divulge. He should seek out background information which may support known facts eg (1) that a clerk accused of falsifying figures to improve his income is already in considerable debt to a loan club, (2) that a fitter's mate accused of stealing copper pipe is engaged in his off duty time in renewing all of the plumbing at his home.

6 Having established the facts at the scene – identified the property – questioned the complainant, witnesses and informants fully the investigator may *now* turn his attention to any person who is accused or suspected and, being in possession of all of the aforementioned information, should tell him in broad terms what is alleged and may question him to establish three things.

a) his identity;
b) whether he considers or complains that there has been a misunderstanding or mistake on anyone's part which has resulted in him being accused or suspected; and
c) so that he is given an opportunity to give any explanation he may wish to offer.

Careful written note should be made at the time of questions put to a suspect or accused person and what he says in reply should be written down *verbatim* without comment or challenge. If the accused gives an explanation which completely exonerates him and to the satisfaction of his accuser, no further action is necessary apart from reporting the facts.

There are many cases in which allegations are met by a blunt denial from accused persons and with a counterclaim that his accusers are liars, swindlers, twisters and every other kind of rogue. From the investigator's standpoint these are merely statements to be recorded showing the accused's reaction and reply and to be reported to some other body ie a senior security officer, top management and/or the police and perhaps eventually a court.

7 If the explanation from the accused is not a complete exoneration or does not satisfy the accuser the investigator must pursue the

matter by oral or written report to his supervisor or to the person who ultimately decides security matters in the organization for the next action to be decided upon. What that decision may be is discussed under the heading of Company Policy in Chapter 2. If there is to be an arrest, the police must be called promptly and the accused should be told of that decision. If there is to be further enquiry occasioning some delay then the suspect should be told of this forthwith.

It is important to remember that the investigating officer is NOT conducting a trial and he may never assume the responsibility of weighing the evidence to decide whether or not the accused is guilty – that is a job for a magistrate or other court of trial. Furthermore, security should curb the old tendency for someone in management to decide that an accused person is guilty of a crime and to sack him from his job without prosecution. That is no longer good practice because employees have successfully claimed through the industrial courts that they have been wrongfully dismissed on the grounds that the original accusation has not been formally proved against them. In some cases reinstatement has been ordered in others substantial compensation has been awarded against the employer whilst in a few cases a civil action for defamation of character has been taken. Management should be advised of the facts known to the investigator and these should include comment upon the attitude of the accused as to whether or not the accused disputes those facts. If that is so management should be advised to have the accusation proved in court before taking disciplianry action. If the facts are not in dispute by the accused then management may take whatever course of action it deems to be appropriate in the circumstances.

Returning to the security officer's investigation, the following matters need further clarification and discussion:

1 The identification of persons
2 The identification of property
3 Describing and preserving the scene of an incident
4 The time and period of loss
5 First steps to be taken at a scene of an incident
6 Questioning witnesses and others
7 Taking written statements

The Identification of Persons

This really means taking a written note of a person's full name, private address and/or a works number, type of employment and department in which employed. If an investigator does not know an individual personally he should ask to see some verification of the information such as a card or document which the person may be carrying bearing some of the details given. Care should be exercised when dealing with a stranger to check that the content of documents produced by him are in fact known by him and do relate to him and it is good practice to accept the wallet or document to examine the contents and then to ask him to tell you what is shown on the document. If it bears a signature such as do driving licences ask him to sign his name on a piece of paper (or a notebook) whilst you hold the licence and then compare the signatures.

It is a growing practice in industrial and commercial houses for identity cards bearing photographs to be worn in a conspicuous position on the clothing of all persons engaged in vulnerable areas and obtaining the identification of such persons means a scrutiny of the pass, recording the pass number, and possibly asking for a signature to compare with the specimen on the pass. If the full practice is applied to everyone there will be no special cause for alarm in the minds of someone under suspicion if he/she receives the same treatment as all the others and it is possible to use special care with some verifications without seeming to do so. If identification cannot be obtained by an exchange of verbal questions and answers a brief description of the individual should be recorded by the security officer who should take every opportunity of asking others to provide clues to the identification of the person in question. A description should include assessments of sex, age, height, build, facial colouring and shape of face, descriptions of nose, mouth, ears and facial hair growth if any, and colour of the eyes. Clothing may be briefly described by reference to outer garments and hat. The description should conclude with a note of any special characteristics such as a slouch or long stride, hip swinging gait, face hooded, scowling expression, whistling, nervous fingering of beads or watchstrap, scratching, possession of bags or equipment, motor car etc.

The experienced and conscientious security officer will train himself to mentally note as many features as possible of every newcomer or stranger he meets and when the call comes for him to exercise his skill he will be elevated in the eyes of others for his efficiency.

The Identification of Property

The black lady's umbrella does not describe the umbrella, it describes the lady in the same way as the brown dog's collar describes the dog when it is the collar we want to identify. A description of property must be clear and capable of interpretation by anyone who reads it. A knife may be clasp, flick, pocket, table, sheath or carving with a blade 'x' centimetres long of polished or unpolished metal or even of plastic material and with a wood, ebony, plastic, metal, bone handle which is plain, smooth, carved, chased or engraved. Some reference must be made to each of those features and to any others which will single that knife out from others of similar type, pattern, design or make. The identity of the knife may be the means of identifying or of proving a case against a particular person and evidence of *that* knife having been seen in the possession of *that* person could be the cornerstone of an otherwise weak case. Therefore, it is vital that the defence should not be able to throw doubt upon the identification of the knife.

It may become necessary to formally prove continuous movement of the knife from the moment it was seen in the hands of the accused, where he threw it, who saw it there, who picked it up, to whom did he hand it, who else saw that transfer, what did this person do with it, where is it now, can he give evidence of having seen that knife in possession of the accused, if so why is he so sure that it is the same knife, if he did not see it in possession of the accused can the man who did now be sure that this is that knife since it has passed out of his possession? Has anyone labelled the knife and initialled the label, if not that must be done now and written statements must be made by all who handled the knife that the knife now labelled is the one which he/she refers to in the statement which describes what he/she saw happen to the knife.

All persons involved may be required to give evidence and asked to say how or why they are sure that that knife (exhibit 1) is the same knife as that to which they testify in their evidence. Laymen may have difficulty in being so sure but a security officer should not because he should have noticed some peculiarity or particular thing about the knife which causes him to distinguish that knife from others.

That simple illustration should be the basis of thinking and action with regard to any item of property which may become an exhibit to be produced in evidence in a criminal trial. The illustration also serves to show how important it is that articles which may become exhibits should pass through as few hands as possible. It is not possible in this chapter to discuss all of the anomolies and mistakes that can be made when describing articles but just to channel the reader's thoughts in the right direction, mention is made of jewellery and the importance and correctness of describing stones as white, red, blue or green and not as diamonds, rubies, sapphires and emeralds because if that does happen a company and the security staff may find themselves challenged for the value of the real thing when only glass or beach stones were in fact present in an item which passed through security hands before restoration to its original owner. Consideration should be given to the possibility of substitution before, during or after the item was in security hands. It is also essential to count the stones and to record the number on every document upon which reference is made to the item – just in case one worth £1000 goes missing!

Describing and Preserving the Scene of an Incident

When referring to the location of an incident, the scene of a crime, or place from which something is missing, the clearest indication should be given of that location in any verbal, written or telephoned communication. 'The cleaners' cupboard on the top floor' does not describe the place nearly so well as 'The cleaners' cupboard on the fourth floor next to room 404'. It is of small value to say 'in the works yard near to the boiler house' when the yard extends on three sides of the boiler house and when the scene would be more accurately described as 'in the works yard 14 feet from the west side of the boiler house and in line with the hydrant on the boiler house wall'. In brief, scenes and locations should be described by reference to surrounding features so that if different persons follow the same description of the scene they should arrive at precisely the same spot.

Preserving the scene refers to any action which can be taken to ensure that there is no interference with or disturbance of a location – that marks of whatever kind are not rubbed or obliterated whether they be fingerprints, splash marks, tyre marks or paint. Protection must be given to other matters such as glass fragments, safe ballast, sawdust and cement dust to ensure that they are not kicked around or walked away on the soles of shoes worn by amateur investigators. Cutting tools and instruments used for leverage always leave traces of their use. Fibres and hairs adhere easily to rough surfaces and can be of value to the professional investigator. It should be remembered that paper is a good carrier of fingerprints especially in an unfolded or uncrumpled state and therefore papers and books on desk tops or in drawers should not be disturbed if the police are being called no matter how urgently the user of the desk wishes to get on with his work.

Preservation of materials is advisable in all serious crime cases regardless of the fact that someone is present who says that he committed the crime because at some later date he may rescind that statement. If too much reliance has been placed upon his first admission of guilt it could happen that vital clues at the scene have not been gathered or protected so that ultimate proof sufficient to convict him is not available at his trial.

At the scenes of accidents a careful first look round should be taken to notice whether safety devices which should be used are being used, that protective clothing is being worn, that guards are on the plant, that power units are correctly connected and are not overloaded, and to note the condition of floors, staircases or other features in the vicinity. This 'taking stock of the situation' is advisable at an early stage because things have a habit of being changed or of moving so that what was previously out of order is now in order. The perpetrator of that change or movement will probably be some 'do gooder' who closes a gate or removes excess equipment from a power supply to make things look better for the victim or, the perpetrator may be a supervisor who was causing (or permitting) work to be done without the proper safeguards and now acts promptly to save himself from criticism. Noticing these things can be of just as much importance to the injured party as to the company and a security officer's impartial comment about his observations can be vital evidence at a subsequent board of enquiry, arbitration panel hearing or court proceedings whether criminal or civil. Preserving the crime scene in the memory can be just as important as physical preservation carried out by roping off the area, by evacuation or by shielding from the weather and prying eyes.

The Time and Period of Loss

Recording time can only accurately be done at the time of and at the scene of an event. If delay occurs so that there has to be an assessment of time based upon memory there will undoubtedly be some discrepancy and the longer the delay the greater margin of error will there be.

This subject touches upon the value and use of notebooks so that there is readily available to a security officer a place upon which to make vital notes such as the jotting down of times. One frequently reads a book or watches a play or film in which an alibi is established by reference to time and it is a fact that in real life no less than in fiction the time factor can be very material to an investigator. When interviewing witnesses and others it is important to take some care in establishing times or periods mentioned by the interviewee as to how or why he fixes times and/or dates and to include his explanation of those matters in his statement. If this witness can testify that an article which is missing was last seen on such a date or time, is there any other person who may have had an opportunity of seeing it after that time and before its loss was noticed – if so that person must also be interviewed. If losses from stock can only be assessed as vaguely occurring over lengthy periods there is something wrong with the system of recording and reviewing stock control and recommendations should be made for improvements in that direction.

First Steps to be Taken at the Scene of an Incident

An incident may be an accident at work involving an employee, a contractor, or a visitor or it may be the discovery of a theft or other crime or the discovery of goods hidden away pending unlawful removal from the premises or it may be a fire or small explosion or a cash office hold-up or a fight between visiting drivers.

Examples of action which should be taken or considered in particular crimes pending arrival of police are as follows:

Wages snatch on site

Close the gates and any other exits. Prevent *any* person from leaving the premises whether staff, customers, contractors or visitors. Preserve the scene.

If suspects have been detained keep them under closest surveillance and if more than one keep them apart.

Wounding

Detain assailant under close surveillance apart from scene of attack and apart from the victim pending removal of that party for medical attention. Render first aid or cause others to do this. Ensure that a weapon is *retained by whoever took it away from the assailant.* In court proceedings it will be necessary to identify this weapon and the fewer hands through which it passes the better because each person who takes possession of it will be called to give evidence for continuity of identification.

Serious Damage

Guard the scene and preserve for fingerprints. Keep others out of danger.

Arson

Guard the scene from inquisitive nobodies and amateur sleuths. Look out for receptacles foreign to the environment. Enquire for witnesses of the early stages of the fire to assist in establishing source and cause.

Attack on a safe

If it is a disturbed attempt – keep away – keep others away, it may be in a highly dangerous condition. Do not touch or permit others to touch light switches or telephones in the vicinity and keep patrol radios and similar electronic equipment at a distance.

Sexual offences

Preserve the scene – allow no change of clothing or visits to a toilet by either complainant or suspect. If toilet visit compulsory the suspect must be closely watched.

Disclosure of fraud, forgery or falsification of accounts

Shepherd informant into top managers office to repeat his story and for a policy decision to be made upon initial action to be taken. Keep informant in that or adjacent office whilst initial action is instigated. In these cases it is better to take no action at all prior to police arrival unless urgency is created by the suspects departure from the premises or otherwise.

In all cases the sequence of security action should be –

1 the preservation of life
2 the protection of remaining property

3 the restoration of good order
4 the preservation of evidence and protection of the scene
5 the communication of information to management
6 the recording of facts, the identification of persons and the taking of written statements.

Obtaining medical aid or the giving of first aid treatment should be the first consideration where necessary. The protection of remaining property involves fire fighting, shutting down valves or plant to prevent danger and wastage of materials and the confiscation of any weapon or instrument by which persons or property may be further injured.

The restoration of good order means that fighting must be stopped or that an offender must be restrained and/or detained as necessary.

The preservation of evidence and protection of the scene and exhibits has already been referred to in this chapter.

The communication of information to management is a matter of importance so that prompt action can be taken by them to bring to the scene the police and specialists such as a safety officer, technical or scientific advisers or an insurance assessor.

Finally, the security officer must be educated and able to report details and to prepare from notes a comprehensive report of all that happened initially, in what way the incident developed, making recommendations as to what measures remain to be taken. Attached to the report should be any written statements taken in connection with the incident or a note that steps are in hand to obtain statements for submission later.

Questioning Witnesses and Others

Before discussing this subject at all there are two misconceptions in the minds of many people which need to be eliminated in order that the Judges' Rules and the guidance which accompanies them may be clearly understood.

Firstly it is wrong to believe that anyone accused of or suspected of a crime must be cautioned before he/she is asked any questions at all.

Secondly it is wrong to believe that immediately a person is told that he/she is being arrested that person must be cautioned. Now please read on.

The Judges Rules give guidance to police officers and others charged with the duty of investigating offences or charging offenders as to the circumstances in which oral and written statements attributed to *accused persons* will be accepted in evidence by the criminal courts. It is very important therefore that security officer's who are charged with the duty of investigating offences or charging offenders should understand the Rules and should comply with them as far as practicable.

A security officer's responsibility in this respect may depend very much upon his job description and in some few companies the duty of investigation may only be invested in those with some seniority. Those cases apart let us start from the basis that any trained security officer whether working alone or as part of a large team is a person charged with investigating offences and is therefore a person to whom Rule VI applies.

A trained security officer should accept that it is part of his or her professional status to know and confidently apply the Judges Rules in appropriate circumstances. They should understand that the Rules are for guidance only and that Judges have an overriding discretion to exclude any evidence which is unduly prejudicial to an accused person or to admit in evidence that which is tendered as a sincere attempt to be fair and impartial. If the guidance given in the Rules has clearly not been complied with relevant evidence of questions and/or answers is liable to be excluded from court proceedings.

The Judges Rules and Administrative Directions to the Police are contained in a document under that title published by HMSO and originating in a letter from the Home Secretary to Chief Constables dated January 1964. In that letter the Home Secretary wrote –

> 'The new Rules differ in certain important respects from the old. It will be observed, in particular, that two forms of caution are prescribed according to the stage which an investigation has reached. One is to be given when an officer has evidence which would afford reasonable grounds for suspecting that a person has committed an offence. After this caution questioning may continue, but a record must be kept of the time and place at which such questioning began and ended and of the persons present. The second form of caution is to be given as soon as a person is charged with or informed that he may be prosecuted for an offence. Thereafter questions relating to the offence can be put only in exceptional cases, where they are necessary for the purpose of preventing or minimising harm or loss to any person or to the public or for clearing up an ambiguity in a previous answer or statement.'

It should be noted by former police officers that the old Rule 3 had gone ie 'Persons in custody should not be questioned without the usual caution being first administered'.

In a preamble to the new version of their Rules the Judges made the following observations.

'These Rules do not affect the principles

a) That citizens have a duty to help a police officer to discover and apprehend offenders;
b) That police officers, otherwise than by arrest, cannot compel

any person against his will to come to or remain in any police station;

c) That every person at any stage of an investigation should be able to communicate and to consult privately with a solicitor. This is so even if he is in custody provided that in such a case no unreasonable delay or hindrance is caused to the processes of investigation or the administration of justice by his doing so;

d) That when a police officer who is making enquiries of any person about an offence has enough evidence to prefer a charge against that person for the offence, he should without delay cause that person to be charged or informed that he may be prosecuted for the offence;

e) That it is a fundamental condition of the admissibility in evidence against any person, equally of any oral answer given by that person to a question put by a police officer and of any statement made by that person, that it shall have been voluntary, in the sense that it has not been obtained from him by fear of prejudice or hope of advantage, exercised or held out by a person in authority, or by oppression.

The principle set out in paragraph (e) above is overriding and applicable in all cases. Within that principle the following Rules are put forward as a guide to police officers conducting investigations. Non-conformity with these Rules may render answers and statements liable to be excluded from evidence in subsequent criminal proceedings.'

The Rules are as follows:

Rule I When a police officer is trying to discover whether, or by whom, an offence has been committed he is entitled to question any person, whether suspected or not, from whom he thinks that useful information may be obtained. This is so whether or not the person in question has been taken into custody so long as he has not been charged with the offence or informed that he may be prosecuted for it.

Rule II As soon as a police officer has evidence which would afford reasonable grounds for suspecting that a person has commited an offence, he shall caution that person or cause him to be cautioned before putting to him any questions, or further questions, relating to that offence.

The caution shall be in the following terms:

> 'You are not obliged to say anything unless you wish to do so but what you say may be put into writing and given in evidence.'

When after being cautioned a person is being questioned, or elects to make a statement, a record shall be kept of the time and place at which any such questioning or statement began and ended and of the persons present.

Rule III

a) Where a person is charged with or informed that he may be prosecuted for an offence he shall be cautioned in the following terms:
   'Do you wish to say anything? You are not obliged to say anything unless you wish to do so but whatever you say will be taken down in writing and may be given in evidence.'

b) It is only in exceptional cases that questions relating to the offence should be put to the accused person after he has been charged or informed that he may be prosecuted. Such questions may be put where they are necessary for the purpose of preventing or minimizing harm or loss to some other person or to the public or for clearing up an ambiguity in a previous answer or statement.

Before any such questions are put the accused should be cautioned in these terms:

> 'I wish to put some questions to you about the offence with which you have been charged (or about the offence for which you may be prosecuted). You are not obliged to answer any of these questions, but if you do the questions and answers will be taken down in writing and may be given in evidence.'

Any questions put and answers given relating to the offence must be contemporaneously recorded in full and the record signed by that person or if he refuses by the interrogating officer.

c) When such a person is being questioned, or elects to make a statement, a record shall be kept of the time and place at which any questioning or statement began and ended and of the persons present.

Rule IV All written statements made after caution shall be taken in the following manner:

a) If a person says that he wants to make a statement he shall be told that it is intended to make a written record of what he says. He shall always be asked whether he wishes to write down himself what he wants to say; if he says that he cannot write or that he would like someone to write it for him, a police officer may offer to write the statement for him. If he accepts the offer the police officer shall, before starting, ask the person making the statement to sign, or make his mark to, the following:

'I, ..., wish to make a statement. I want someone to write down what I say. I have been told that I need not say anything unless I wish to do so and that whatever I say may be given in evidence.'

b) Any person writing his own statement shall be allowed to do so without any prompting as distinct from indicating to him what matters are material.

c) The person making the statement, if he is going to write it himself, shall be asked to write out and sign before writing what he wants to say, the following:

'I make this statement of my own free will. I have been told that I need not say anything unless I wish to do so and that whatever I say may be given in evidence.'

d) Whenever a police officer writes the statement, he shall take down the exact words spoken by the person making the statement, without putting any questions other than such as may be needed to make the statement coherent, intelligible and relevant to the material matters; he shall not prompt him.

e) When the writing of a statement by a police officer is finished the person making it shall be asked to read it and to make any corrections, alterations or additions he wishes. When he has finished reading it he shall be asked to write and sign or make his mark on the following Certificate at the end of the statement:

'I have read the above statement and I have been told that I can correct, alter or add anything I wish. This statement is true. I have made it of my own free will.'

f) If the person who has made a statement refuses to read it or to write the above mentioned Certificate at the end of it or to sign it, the senior police officer present shall record on the statement itself and in the presence of the person making it, what has happened. If the person making the statement cannot read, or refuses to read it, the officer who has taken it down shall read it over to him and ask him whether he would like to correct, alter or add anything and to put his signature or make his mark at the end. The police officer shall then certify on the statement itself what he has done.

Rule V If at any time after a person has been charged with, or has been informed that he may be prosecuted for an offence a police officer wishes to bring to the notice of that person any written statement made by another person who in respect of the same offence has also been charged or informed that he may be prosecuted, he shall hand to that person a true copy of such written statement, but nothing shall be said or done to invite any reply or comment. If that person says that he would like to make a statement in reply, or starts to say something, he shall at once be cautioned or further cautioned as prescribed by Rule III(a).

Rule VI Persons other than police officers charged with the duty of investigating offences or charging offenders shall, so far as may be practicable, comply with these Rules.

Rule I clears the way for any person to be asked questions which will progress an enquiry into a complaint or allegation of criminal activity. All questioning should be carried out in a forthright and commonsense way and with the sole object of establishing the truth about the matter under review. The investigator should always try to be fair to the person who is being questioned and should scrupulously avoid any method which could be regarded as unfair or oppresive. If all persons suspected or not are treated in the same way there can be little cause for complaint or criticism of the investigator's actions.

The truth if established, may not be what the questioner anticipated or hoped for but it should give an accurate picture of the facts so that a correct course of action may follow. One hard fact to bear in mind is that it is human nature to place all of the blame upon someone who is not present at the interview. There is much truth in the saying – a good listener will learn from others – a chatterer learns only from himself. All questions should have an aim and purpose and generally should follow a prepared pattern or natural sequence of events which stems from logical thought.

The questioner must not be diverted from his/her aim and must be firm with those who insist upon rambling away from the point at issue. The witnesses of an incident may be reluctant to say what they know out of misplaced loyalties or a desire not to become involved and with such persons great patience is sometimes very necessary.

The questioning of complainants to obtain detail and verification of their first complaint and the questioning of witnesses and others who may assist an enquiry should, in the majority of cases, be conducted out of the hearing of an accused or suspected person. The investigator should obtain as much information as possible from complainants, witnesses and others and should then –

- i identify the suspect;
- ii ensure that the suspect is not alleging mistaken identity or other error on the part of his accusers; *and*
- iii ensure that the suspect is given the opportunity of offering an explanation of the circumstances;

the officer should then make his report to higher authority.

By following that pattern of questioning the officer does not have to be involved with the rest of the Judges Rules because the guidance of Rule II reads:

> As soon as a police officer (security officer) has evidence which would afford reasonable grounds for suspecting that a person has committed an offence, he shall caution that person or cause him to be cautioned *before putting to him any questions, or further questions, relating to that offence.*

In the case of Walmsley v Young 1974 Criminal Law Review 548 it was held that before the Rule II caution becomes obligatory there must be sufficient evidence to show a prima facie case against a person under suspicion. Therefore in practice a security officer is not in a position to say that there is 'evidence which would afford reasonable grounds for suspecting that a person has committed an offence' unless the officer

a) saw the offence committed (in which case there is little need to question anyone); *or*
b) has listened to all of the story and answers to his questions put to informants, witnesses and complainants; *and*
c) has listened to the answers put to the suspect to clear up any suggestion of mistake together with his explanation,

and if at that stage a subordinate security officer does not intend to ask the suspect *further* questions there is no reason to give a caution and this can best be left to a more senior officer or police officer when that person has carried out *his* investigation. However, if the subordinate security officer approaches a suspect who forthwith or during the interview, says something which amounts to an admission of guilt the security officer should record what is said and then give the caution quoted in Rule II. Under those circumstances the confession would be regarded as voluntarily given and would be admissible in evidence.

It is a misunderstanding amongst many security people that when they make up their minds that they will not release the suspect without further enquiries that they must give a caution but Rule II should be understood to mean what it says which is that it is necessary to caution at that stage 'before putting any questions, or further questions relating to that offence'. If it is intended at that stage to pass the enquiry to someone else and not to put to him or her any further questions there is no obligation to give a caution because the accused is not being required to answer questions the answers to which will be given in evidence. If the accused is told at that stage that he/she will be detained to await the police or pending a decision by the works director questions are not being asked and the caution is not necessary. If at that stage the accused says something to imply his/her guilt of an offence what is said should be recorded without question and a caution as in Rule II should then be given.

A senior security officer coming into an enquiry at any stage should conduct his or her independent investigation and must comply with

Rules I and II as and when the progress of the enquiry dictates correct action in accordance with the Rules.

It is extremely unlikely that a company, with or without legal advice, will decide upon a private prosecution without the aid of the police but in the event of such a rare circumstance it is obligatory for the Rule III caution to be given during any interview in which the person being prosecuted is informed that proceedings are being taken. If police are being involved the time for the Rule III caution is at the time the accused is told *by them* that he is charged or summoned for an offence.

Rule IV relates to the taking of written statements from accused persons who have been cautioned and can seldom have an application to the work of subordinate security personnel. If however accused persons indicate that they wish to have a voluntary statement recorded either by writing it themselves or by dictation, senior security officers and investigators must know and comply with Rule IV. An employee against whom criminal proceedings are pending may hand to the company investigator a written confession prepared at his home address and it would be proper for the investigator to ask the employee to write a statement as outlined in paragraph (c) of Rule IV and then to ask him to make reference to his letter which should be attached as an original document to such statement.

In short a person under suspicion or accused should not be deprived of the right to make an oral or written statement if he/she wishes to do so and any security officer should be prepared to accept or to assist in recording such a statement in compliance with Rule IV.

Rule V relates to the exchange of statements between persons who are jointly accused of offences and after they have been charged or informed that they will be prosecuted and that situation can have no application to the work of security personnel.

Appendix B to the Home Office memorandum on the Judges Rules is headed 'Administrative Directions on Interrogation and the Taking of Statements'. Broadly these direct that questioning shall take place with all parties present being seated – that arrangements are made for refreshment intervals – that this may not include alcohol – that what is said should not be translated into 'official' vocabulary – that an interpreter should record a statement in the language spoken for translation later – that children and young persons should only be questioned in the presence of their parent, guardian or an adult not a police officer and of the same sex as the child or young person – that when they are formally charged, a written notice detailing the charge or offence be handed to persons who have been arrested without a warrant – that persons in custody should be allowed to telephone a solicitor or friend 'provided that no hindrance is reasonably likely to be caused to the processes of investigation, or the administration of justice by his doing so'. A person in custody also should be given facilities for sending letters or telegrams at his/her own expense.

Before leaving the Judges Rules the reader's attention is called once again to the wording of Rule I which authorizes questions to be put to any person suspected or not from whom useful information may be obtained whether or not he is in custody so long as he has not been charged with or informed that he will be prosecuted for the offence for which he was arrested. Therefore detaining a suspect (refusing him leave to depart of his own free will) and then continuing to question without giving a caution is quite in order – it is only when the investigator has sufficient evidence upon which to base a prosecution or charge that he/she should either stop asking questions or administer the Rule II caution. Rule I also authorizes questions to be put to persons in custody about any other offence of which he may or may not be suspected.

For example, it would be perfectly in order for a security officer to arrest for theft a man who has been stopped when leaving company premises in possession of a litre of paint stolen from the workshop and then, because the officer knows that a wallet containing two £5 notes has recently been stolen from clothing in the workshop, to question the man about his possession of two £5 notes carried loose in his jacket side pocket. A further example would be that a police officer who has arrested a youth for taking a motor-cycle without authority may question without caution the youth about his route of travel, manner of driving and possible involvement in a fatal accident just reported to have taken place in the area.

By applying that kind of reasoning it will be clear that investigatory work is not impeded by the Rules particularly in those early stages of an enquiry when security officers are likely to be involved.

Taking written statements

Taking down a statement of a witness, informant or complainant in writing is a duty which may be carried out by some security officers far more frequently than others or may be carried out by officers with considerable experience of statement taking. Those less practised may require guidance and they may find the following notes to be helpful. The purposes of taking a statement down in writing are:

a) to record what a person may say whilst the matter is fresh in his or her memory and as a permanent record;
b) to provide a basis for further enquiries and cross checking of information;
c) to provide information upon which further action may be taken or to confirm grounds for any action which has already been taken.

It is an advantage if one person takes the statements from several persons in the same case so that gaps in evidence may be filled or highlighted, variations or discrepancies noted and cleared up and for important aspects to be reiterated for corroboration or emphasis. If a statement refers to lists of property or cash items it is advisable for the witness to prepare personally that list or extract from a record and then to make reference to the list in his statement to which the list is attached so that the list may become an original exhibit. Statements should if possible be written on plain paper but in rare circumstances a notebook may be used.

The taking of a statement should be unhurried so that the person making it has an opportunity to collect his or her thoughts into a logical sequence of events. He or she may write the statement if they wish or may dictate what they have to say for a security officer to do the writing. The person taking the statement should not unduly interfere with what is being said but may keep the statement to the main point at issue and should clear up ambiguity in what is being written as the statement progresses. Upon completion the statement should be read over by or to the person making it and that person should have an opportunity of correcting anything which is considered to be incorrectly recorded. They should then be asked to sign the foot of each page of the statement and those signatures should be countersigned by the person who has taken the statement or caused it to be written and he should then add the time and date of signing.

An example of a statement well taken would be as follows:

I am Lydia Phyllis Wells, aged 40 of 33 Stanfield Close, Bryley, Yorks, a typist employed in the Sales Office of John Willis & Co Ltd. Timber Merchants of Bryley. On Monday 3 November 1975 I was working at my office desk which is in the Sales Office on the first floor and overlooking the works yard. At about 12.5 pm I looked out of the window towards a road trailer where a gang of men were unloading long timber by means of two mobile cranes. One crane was supporting each end of the timber which had been lifted about four feet above the trailer. At that time Joseph Woolard, one of the gang, was standing on the trailer giving hand signals to the crane operators. Suddenly something seemed to slip on the crane hoist furthermost from my window and that end of the timber dropped about two feet and remained in that position. As the timber fell it knocked against Mr Woolard on his left side and he fell off the trailer to the ground.

I am quite sure that the timber did strike Mr Woolard and that he did not just jump off the trailer, he was knocked off.

My office window is about thirty feet from where the back end of the trailer was standing at the time and about fifty feet from where Mr Woolard was hit.

Lydia Wells

Statement taken and written at Mrs Wells dictation by J Grafton, Senior Security Officer and completed at 3.30 pm on 3 November 1975 in the Personnel Department.

John Grafton

That statement is simple but direct. It contains all relevant facts which this witness knows in a good sequence and presents a clear word picture of the event. The witness was obviously asked two important questions at the end of the narrative which have a bearing upon the impartial enquiry which the senior security officer has been asked by his employer to make.

The statement fills all of the purposes of a written statement in that it records facts known to the witness whilst those facts are fresh in her memory; it provides a basis for further enquiry quite obviously as to why the timber suddenly fell at one end and for enquiry about the method employed to unload long timber; and it provides information upon which responsibility may be assessed for the purpose of compensation to an injured man.

The skills of questioning and of statement taking come naturally to only a very few police and security officers and can be acquired only by years of practice and experience. To assist newcomers to the profession towards an understanding of that skill the author of this chapter strongly supports and recommends the excellent advice contained in Chapter 11 of the Gower Press publication *Practical Security in Commerce and Industry* by Oliver and Wilson who have explained the subject in language which is as clear as it is concise.

The ability to make an impartial investigation and record information are talents which management should recognize in those members of their security staff who possess them and those talents should not be wasted by confining them to the petty accident or occasional employee pilfering.

Attention is called to the comment made under the topic of company policy that the company's right to agree or not to the calling of civilian police to an alleged crime relates only to crimes which affect company owned premises or property. Should the loser of property be an employee or a contractor on site that loser has the right to call police to investigate his loss when that loss may be attributable to crime.

In many cases losers are content that an internal investigation be made by the security staff and it is in those cases that security may properly advise management whether or not they consider it advisable for police to be called in and to call police on behalf of the losers. Similar considerations apply in cases of unlawful damage to privately owned property and to cases of assault upon the person.

In the course of their duties police sometimes have cause to enquire at business premises on matters of routine enquiry such as to trace ownership or usage of a motor vehicle at a particular time and place or to trace a fugitive or to interview a witness or suspect of an incident which has taken place elsewhere.

The reception of police on such enquiries differs widely in commercial and industrial houses and, whilst they may be received with tolerance for enquiries concerning the firm's own vehicles or drivers, their reception may not be so cordial and their enquiries may be regarded as a waste of company time when police wish to interview employees on matters not involving the company. It is the task of security to liaise with police with the best of good feeling at all times and where it is known that there is company objection to on-site enquiries which could be made when employees are not working that fact should be tactfully explained to visiting police officers. It is a feeling with some employees that employers should not give facilities for police interview whilst they are working and in the interests of industrial relations some employers accept that point of view. Management should be advised by security when the support of such an objection amounts to an obstruction of police.

In concluding this chapter the reader is reminded of the paragraph about asking questions with which the chapter commenced and in particular to the phrase 'in the right or wrong way'. A successful investigation can so much depend upon the personality of the investigator and volumes could be written about tact, discretion, firmness, bonhomie, persuasion, humour and aggression all of which attitudes have their right place and proper use when contact is made between humans. A quiet, composed and polite approach to everyone is essential for success and if that is received by the reaction 'You can only do your job officer' the investigation has made a very good start.

Adopting a pose of authority which infers that persons being spoken to are under duress or obligation to co-operate is an attitude least likely to succeed and security officers should never try to use 'official' language which is a characteristic of stage detective plays. People like to be treated as sensible understanding people and whilst it may take a little longer and require more patience to obtain the comprehension of some more than others no one should be made to feel that they are under ridicule or that the interviewer is 'talking down' to them.

In a contra-situation an investigator should never have an inferiority complex in that he is at a disadvantage by reason of status, education or intelligence. The best antidote to such a complex is to have confidence in one's knowledge of one's job and to do that job without fear or favour, being firm in one's resolve to elucidate all of the facts whilst retaining a sense of fair play and complete detachment from personal involvement.

The Investigation of Fraud

Home Office Reports and the reports of Chief Police Officers say that crimes by fraud are to some extent taking the place of crimes by physical force and therefore this Chapter cannot be complete without reference to some of the special features of fraud investigation.

Chapter 16 contains a full quote from Sections 15, 16 and 17 Theft Act 1968 because those Sections cover such a wide range of fraudulent conduct that there is likely to be found an offence to fit the circumstances of fraudulent conduct coming to the notice of any security man at any time. There are other Sections of Theft Act dealing with the accounting liabilities of directors and officers of companies and there are a number of other Acts of Parliament dealing with specific types of fraud such as Hire Purchase Act 1965, Companies Act 1948 and The Prevention of Fraud (Investments) Act 1958. Such crimes by or against companies are for investigation by specialists and a limitation upon space requires the authors of this book to confine their attention to advice and guidance for dealing with alleged fraudulent actions 'within' commerce and industry ie internally by staff and others. The erosion of a business by small defalcations can seriously affect profitability and it only requires there to be one or two regular operators in that direction to bring a small business to bankruptcy. The dishonest shop assistant amidst a staff of three can drain £20 per week from the takings by undertilling and similar short-changing' actions and there may be difficulty in identifying which of the three assistants it is unless the shop owner goes to the expense of employing an outside contractor to carry out test purchasing in his shop.

A foreman for a contract painting company took on extra contracts on his own account and directed his employer's men to do his own work in the afternoons. The men thought that they were being switched about in their employer's jobs in order to cover up delays in completing contracts. This fraudulent use of manpower was disclosed when a painter was involved in an accident at one of the foreman's jobs.

A cash office clerk inserted into the accounting system of his employer forged claims for travelling and incidental expenses in the name of a fictitious company representative and was able to draw the money by production of the approved claims in his own department. This he would not have been able to do if the item had been payable through the salaries system.

This small sample of cases illustrates the opportunities for crime which are presented to dishonest employees by weak and uncontrolled administrative systems – they also demonstrate the value of administrative systems being (a) prepared or approved by someone with a security outlook who can see the loopholes; or (b) supervised by an audit examination which is designed to reveal weaknesses.

So many companies continue to work with operating and administrative systems which are very much out of date and out of keeping with the modern trends in human behaviour. They are systems which in the old days were adequate and which relied upon the trustworthiness and loyalty of the work people but today they merely provide opportunities for dishonesty and a temptation which more and more employees have not the will to resist.

Mention might appropriately be made here of the fraudulent activity which is perhaps better known by the old name of embezzlement – the unlawful retention by a servant of money or goods received on behalf of his master. Today this crime is charged as theft contrary to Section 1 Theft Act. Poor accounting systems have in the past encouraged trusted and long serving employees to fall for the temptation to embezzle and to falsify books to conceal their crimes. A safeguard against such conduct should be for a periodical audit of the work, and review of the system, by head office or outside auditors. An accounting system may have been devised or varied over the years by the very people who operate it and although there may be no grounds for mistrust of the present operators an audit may reveal weaknesses in the system which would be a temptation to a replacement operator following upon the retirement of a good and faithful servant.

Frauds need to be investigated very carefully and for this reason many firms prefer to employ persons with police investigating experience as their senior security people. Ex-policemen have by virtue of their training and experience a certain 'know-how' for recognising symptoms and hints of evidence of fraud. They know what use to make of such evidence and how important it is to preserve that evidence and any other which they may unearth with the utmost confidentiality. There is no mystery about acquiring the policeman's know-how – it is obtained by an intelligent and thorough study of the law coupled with concentrated attention to the examination of detail. Facts are of paramount importance and all ambiguity must be challenged and clarified. Above all there must be a clear recognition that collusion between fraudsmen or women and persons holding positions of trust may exist at any level of a business and therefore there is a need to carry out investigations with the utmost delicacy and tact in order to avoid even a hint of suspicion becoming known to those involved.

There are occasions when persons who keep records should not become aware at least in the early stages of an enquiry that those records are the subject of examination or suspicion and there are other occasions when it is necessary and correct action to descend upon such a person at the outset of an enquiry to seize his books and records or to ask for his keys to gain access to such documentation. A decision as to which is the correct course of action must be made upon receipt of, and be based upon, the initial information presented to the investigator.

Another aspect of the work is that the investigator should endeavour not to disclose an unusual interest in a particular person, or section or department, until he is ready to take forthright action. In this respect the rare presence of the security man in a department manager's office could put shady persons 'on their guard' – either the security man must go into that office as a daily routine so that no visit is any more alarming than another or the manager must meet the security man by clandestine arrangement elsewhere on or off the site.

The initial steps to be taken in any investigation may depend very largely upon the source of the information and upon company policy with regard to interviewing and interrogation of employees or of automatic involvement of unions in any enquiry affecting union members and must be governed by the company's overall attitude towards prosecution. Therefore the first steps may be an interview with top management to clarify an unsettled policy before any investigatory action is taken at all. On the other hand delay may cause the loss of important evidence because a document which affords a valuable link in a chain of evidence and from which details have been copied onto computer records is likely to be destroyed by shredding as a normal routine and it is the *original* document that will be required for evidence and not a copy. Policy decisions are important because fraudulent activity may also involve the crimes of forgery, bribery and corrupt practice. The best advice to management is for police to be called at a very early stage in the enquiry so that there is no opportunity for pre-arranged explanations, denials and alibis.

Some criminals adopt the belief that the best defence is attack and will promptly allege the corrupt involvement of an employee or security man who is bold enough to accuse them or to question their behaviour. For this reason the investigator should advisedly be accompanied by a colleague or subordinate security officer when interviewing a suspect. If the interviewee is a woman then the investigator should have a woman with him.

With these thoughts in mind the investigator must follow a special pattern of action in cases which relate to allegations or disclosures that there has been a fraudulent alteration of book entries or that monies have been diverted from company to private ends, or that club accounts have been the subject of falsification. In these and similar cases the first steps may need to be more circumspect than those taken in other straightforward theft cases where open investigation of a disclosed or alleged crime is the proper course of action.

Company policy as discussed in Chapter 2 can have an importance here because the following courses of action are open to the investigator:

1 immediate confrontation of the accused with the known facts and an explanation demanded;

2 an open seizure of books or documents for close scrutiny without disclosing to anyone the reason for the seizure;

3 a discreet inspection of books or documents without the knowledge of anyone that an inspection is taking place;

4 a period of inactivity in order to allow time for the development of a fraud or for the intended purpose of falsification to become more apparent;

5 a period of observation to establish who else, if anyone, is involved in this or in a similar fraudulent activity.

Of these courses 2 is only suitable when no particular person is suspected or when the person alleged to be involved is absent or otherwise unlikely to become aware of what is happening or where an immediate checking of facts is essential for the good of the company and without an intention to deal with anyone as an offender.

Course 3 may be necessary because the initial information is of a weak nature or because of the status in the company of the suspect. If 3 is decided upon then inspection must either be made out of office hours by using duplicate keys or during office hours by employing an auditor to call for books etc as a matter of normal routine.

Course 4 would by an unsavoury course of action to most managements and would not be received favourably by the courts if investigation resulted in prosecution. It is too closely related to an 'agent provacateur' situation to commend it.

Course 5 is acceptable provided that company policy does not object to a multiplicity of accusations which might be avoided by making an example of one offender.

Whichever course of action is to be taken it is advisable for the security officer charged with the investigation to have a clear understanding with his management as to whether and at what stage he may call police or to conduct his investigation in full consultation with his manager who may decide at any particular stage of the enquiry to authorize the calling of police.

Nothing travels faster than bad news and one whisper of information to an indiscreet person can 'blow' a whole case.

One other small matter to which some importance may be given in the interests of crime prevention is that cheque books and credit cards very much lend themselves to crime by fraudsmen and should therefore be given extra protection and this means educating every female holder of a bank account about the opportunity for crime which she provides whenever she leaves her handbag beside her desk when leaving her office for an enquiry or to take dictation notes; and every male account holder on every occasion when leaving his jacket containing his wallet and documents unattended on the back of his chair, in a cloakroom, in restaurants, in sports clubhouses, in railway compartments and in the myriad other places where men discard their jackets in summer months or in overheated buildings. A crime prevented is one less to be investigated and that is good cost effective security.

A detailed and excellent work entited *The Investigation of Fraud* by D Campbell published by Barry Rose (Publishers) Ltd jointly with the *Police Journal* from Chichester, Sussex, makes interesting reading.

CHAPTER 19

Arrest of Offenders

'I arrest you . . .'. How many private citizens, and that expression means people other than policemen, possess the necessary knowledge of the law and the courage to speak those formidable words to another private citizen? The number is probably extremely small and yet, in the name of the people and as recently as 1967, Parliament gave its blessing to the Criminal Law Act which, in effect, says that in given circumstances any person may arrest without warrant any person.

Industrial Security Officers are in a special kind of job which they are required to perform without the privileges and protection accorded to their contemporaries the regular police and special constabulary. They are required and entrusted, by their employers to protect that employer's property against intrusion, theft and damage, and to generally preserve the peace with no more powers than those which they possess as private citizens. How important it is therefore that the Industrial Security Officer should fully understand his authority within the law and to have the courage to exercise that authority.

An arrest is the taking or restraint of an individual from his liberty in order that he shall be forthcoming to answer in a court of law for an alleged crime or offence. We are told that it is not necessary to touch or to lay hold upon a person in order to arrest him. Telling him 'I arrest you . . .' means that he is detained or in custody and thereafter he can be guilty of escaping from lawful custody. We are concerned here only with arrest without warrant because it is unlikely that a security officer would apply for and obtain a warrant of arrest without co-operation and liaison with Police.

Until the passing of the Criminal Law Act 1967 the power of a private person to arrest another without a warrant was mainly derived from Common Law and was related to the classifying of punishable wrongs into felonies and misdemeanours.

The Criminal Law Act 1967 abolished the distinction between felonies and misdemeanours and all punishable wrongs became known as offences. The new Act also introduced the term ARRESTABLE OFFENCE as being those offences for which the sentence of a Court may be five years imprisonment or more, and all attempts to commit such crimes or offences.

Crimes which come under the heading of ARRESTABLE OFFENCES include all forms of theft (including handling stolen goods); serious assault; serious fraud and forgery; criminal damage including arson; and all attempts to commit such crimes.

> Section 2(2) of the Criminal Law Act 1967 enacts that 'any person may arrest without warrant anyone who is, or whom he with reasonable cause, suspects to be, in the act of committing an ARRESTABLE OFFENCE'. Section 2(3) of the same Act provides that 'where an ARRESTABLE OFFENCE has been committed, any person may arrest without warrant, anyone who is, or whom he, with reasonable cause, suspects to be, guilty of the offence'.
> Section 3 of the Act provides that 'a person may use such force as is reasonable in the circumstances in the prevention of a crime, or in effecting or assisting in the lawful arrest of offenders or suspected offenders, or of persons unlawfully at large'.

Additional to the above certain Acts of Parliament provide powers of arrest for private persons to arrest persons found to be committing lesser crime than those which fall into the above category of arrestable offences. For example:

1 Section 12(3) of the Theft Act 1968 applies the power of arrest contained in Section 2 of the Criminal Law Act 1967 to the offence of taking or using a motor vehicle or other conveyance without authority, or of attempting to do so. For this offence the Theft Act provides a maximum penalty of three years imprisonment. (See Chapter 16.)

2 Section 25(4) of the Theft Act 1968 provides a power of arrest without warrant to any person to arrest any person who is, or whom he suspects, with reasonable cause, to be committing, an offence of 'going equipped to steal' ie 'When not in his place of abode he has with him any article for use in the course of, or in connection with, any burglary, theft or cheat'. The maximum penalty for a conviction for an offence under Section 25 is three years imprisonment. (See Chapter 16.)

3 Section 4 Vagrancy Act 1824 provides that any person may arrest any person who is found to be in an enclosed yard, garden or area or in any dwelling house, warehouse, coach house, stable or outhouse for an unlawful purpose. The unlawful purpose must be an intention to commit a criminal offence. The maximum penalty on first conviction is three months imprisonment or a fine not exceeding £25. (See Chapter 17.)

These are but a few of the special provisions in the law allowing arrest without warrant by private persons. There are many others which have no bearing upon industrial or commercial security.

The Common Law still provides an authority for anyone to arrest another without warrant to put a stop to an actual breach of the peace that is taking place before him and this of course relates to persons openly fighting or taking part in riotous behaviour.

The crux of the whole subject lies in knowing which crimes are serious enough to come within the term 'arrestable offence' or are specially mentioned in a law which provides a power of arrest to private persons. There is no easy way to acquire this knowledge and it is necessary to study the various criminal wrongs, the methods of committing them, and the powers of the Courts when dealing with persons brought before them for committing those wrongs.

Quite understandably senior management in industry is apprehensive as to the possible repercussions upon the company, and upon individual security officers, in the event of a person being accused, detained and possibly forcibly detained without sufficient grounds for so doing. In order to allay some of that apprehension it may be helpful to set down guide lines for the information of security personnel.

1 When observing the actions of some person who is behaving in a suspicious manner it is much better for a security officer to delay intervention until he is quite sure of his facts, and is in a position to be able to challenge a person by telling what he has seen him do, rather than to jump in too soon only to find himself in a position where it can be said that the officer is falsely accusing the person or trying to make a crime out of an innocent action. Similarly, if the officer is given information by a third party it is better to verify that information, before challenging an alleged offender than to go in with only half a story and then try to rely upon a guilty explanation from the accused.

   A security officer who has taken such measures to verify information or to reassure himself of the facts can have confidence in the knowledge that a theft or other serious crime has been committed and that there is good cause to believe that the person he is accusing or who is accused by others has committed that crime. He should then act with firmness and should not be put off by a retaliatory attitude on the part of the accused.

2 Let us go over Section 2(2) Criminal Law Act 1967 again in detail. 'Any person may arrest without warrant anyone who is *in the act* of committing an *arrestable offence*'.

   If an officer sees someone stealing or finds them in possession of stolen property or inside a building attempting to steal or attempting to commit deliberate damage, or if his attention is called to someone committing a serious assault or setting fire to premises or property he may arrest that person using only sufficient force as may be necessary to prevent the person escaping and *should forthwith tell the accused what he has been seen to do and therefore why he is being arrested.*

Section 2(2) also says: 'Any person may arrest without warrant anyone whom he *with reasonable cause suspects to be in the act of committing an arrestable offence*'.

If an officer sees someone engaged in some activity which can have no other explanation than that he is stealing or committing burglary or attempting to commit serious damage or to set fire to premises or property then he may arrest that person in the manner and form shown above because he reasonably suspects him to be in the act of committing a crime which is an arrestable offence.

Section 2(3) of the Criminal Law Act 1967 says: 'Where an *arrestable offence* has been committed any person may arrest without warrant anyone who is guilty of the offence'.

This subsection provides for an arrest to be made in circumstances in which the offender is not caught in the act of committing a crime but when a security officer knows that a crime has been committed by someone and then finds a person who admits responsibility for that crime.

Section 2(3) also says: 'Where an *arrestable offence* has been committed any person may arrest without warrant anyone whom he with reasonable cause suspects to be guilty of the offence'.

Under this subsection authority to arrest is given to a security officer who knows that a crime has been committed and who then finds, or is called to, a person who is in possession of property from that crime or whose conduct leaves little room for doubt that this is the person who committed that crime.

To summarize those points it must be clearly remembered that the authority to arrest without warrant is given by Section 2 to a security officer –

- i who catches someone in the act of committing an *arrestable offence*; *or*
- ii who is called by a third person to someone who is actually committing an *arrestable offence*; *or*
- iii who catches or is called to someone behaving so suspiciously that there is little doubt that he is in the act of committing an *arrestable offence*; *or*
- iv who knows that an *arrestable offence* has been committed by someone and who then finds, or is called to, someone who admits to that crime; *or*

v who knows that an *arrestable offence* has been committed and who then has good cause on reasonable grounds to suspect that a person he is confronting committed that offence even though he will not admit to it.

3 If there has not been time for preliminary observation or verification of information an officer may yet detain a person against whom an allegation of crime is being made and who appears to be about to abscond. The officer may then thoroughly explore the grounds for the accusation either in the presence of the accused or otherwise and may conclude by asking the accused to give any explanation.

4 These instructions should be carefully read in conjunction with Chapter 18 which deals with the investigation of crime and other incidents where stress is given for the need to concentrate upon obtaining evidence from other employees, losers of property, or witnesses of an incident and to leave interrogation of an accused person to the last. Doubtless, the accused will want to have lots to say and the more guilty he is the more likely it is that his protests will be vehement in order to make his case heard. It must always be remembered that the onus of proving a criminal case rests upon the prosecution and this starts with a full and efficient investigation at the outset to obtain all the evidence available and without relying upon a confession.

Great heed should be taken of the instruction contained in Chapter 17 upon the procedure for dealing with trespass and those suspiciously upon company premises.

5 An arrest should be made as quietly as possible and unnecessary force avoided. An accused person must be treated with consideration. He should be detained in the security office or other convenient place until the arrival of the head of the security department or the police if they have been called. The person should be told why he is being arrested and that the police have been called if that is the case. Once under detention the accused must not be lost sight of even for a moment, or under any pretext. If he asks to go to the lavatory he must be accompanied and watched. A woman will be escorted by a woman. No opportunity must be given for an accused person to dispose of stolen property by hiding it in a lavatory cistern or down the back of a car seat etc. Upon an arrest being made the property in the case will be retained by the officer making the arrest who will hand it to his senior officer or to the police when they arrive.

6 A security officer who challenges a person because of what he has seen that person do, or because of some article that the person has in his or her possession, needs to act with tact and discretion so that

there can be no grounds for complaint subsequently from the accused that he was assaulted or otherwise mishandled. These are frequently side issues which accused persons try to introduce to divert attention from their own guilty conduct.

However quiet and tactful an officer may be in challenging a trespasser or wrongdoer there could arise a situation in which the person spoken to becomes violent and assaults the officer. Should this happen the officer may not retaliate and thus become involved in a brawl but should endeavour to restrain the other against further violence by taking hold of him until he has calmed down or until assistant arrives. In every such case consideration should be given to possible action through a court against the assailant according to the severity of the assault which he has committed. This consideration should be independent of any other proceedings in which the trespasser or wrongdoer may have become involved.

7 The fact that the law allows arrest without warrant for certain offences does not necessarily mean that an arrest must be made in all cases. The power to arrest should be regarded as an authority and not an obligation and should usually be exercised –

i to put a stop to crime actually being committed,
ii to obtain evidence in support of other factors about a known or suspected crime, *or*
iii to ensure that an unknown suspect does not abscond.

The authority contained in the law for private persons to detain or arrest others applies anywhere. Security officers should be aware of this and should recognize that their actions are not limited only to within their company's boundaries. Nevertheless the level headed security man will exercise discretion and act only when action is justified in the course of his employment.

8 The special attention of readers is directed to Chapter 18 where detailed instruction is given about the proper application of the Judges Rules for questioning accused persons.

9 To detain a person is to arrest that person. If a point is reached when a security officer places a restraint upon the movements of another by refusing to allow him to proceed or to leave the room or place where he is the officer is arresting that person and should be in a position to justify that action. It is wrong to believe that an arrest does not take place until a decision to prosecute has been made. Arrest and prosecution are separate and distinct courses of action. There may be an instance in which a security officer is justified in detaining a person but a manager may authorize his release without

> reference to police. Such cases do not make the arrest of the accused unlawful and should not reflect adversely upon the officer because the law allows for the company to exercise discretion as to whether or not they should prosecute in certain circumstances or to take proceedings against an offender otherwise than by preferring charges. Provided that a security officer has acted in good faith and upon sound facts from what he has personally seen or heard from others the detention of a wrongdoer, or even of a suspected wrongdoer, will withstand any subsequent attempt to belittle his action and the company should not be prejudiced by it. (The decision to prosecute is discussed in Chapter 2.)

To summarize: industrial security officers need training in order to acquire that essential ingredient – 'knowledge' – to inspire their confidence; they need to act with restraint and discretion so that they do not blunder into a situation only half-informed and unable to substantiate their action; they need to understand that they must eliminate from their minds any feelings of personal malice towards a wrongdoer and they must handle that wrongdoer in such a way that their own conduct cannot be challenged; they need to understand the Judges Rules and the law of evidence so that they may assist the courts in reaching a just and true finding upon a case before them.

The common law confers upon every adult the duties of citizenship to keep the peace, obey the law and to ensure that others do so. Industrial security officers are ordinary people who have a special place in modern society for they are particularly charged with the duties of citizenship to keep the peace, obey the law and to ensure that others do so and when an officer says to a person 'I arrest you . . .' he is indeed carrying out that duty.

CHAPTER 20

Court Procedure and Evidence

Having taken all of the correct action at the scene of an incident or having successfully handled the initial stages of a complicated situation a security officer may next find that he is called upon to relate what he knows about the matter to any or several of the following:

The head of his department
the works director
a police officer called on site
another police officer at a police station
a magistrate or chairman of the Justices at a Magistrate's Court
a solicitor acting on behalf of his employers or of some other party
a Judge at a Crown Court
a factory inspector
an official tribunal or board of enquiry
a compensation board
a County Court Judge
a High Court Judge

Alternative to that list there might be interviews with private enquiry officers acting for insurance companies or with solicitors and barristers representing each or all of the parties involved in proceedings arising from an incident which the officer has attended and reported upon.

Vital issues may be argued based upon a statement recorded by the officer in his notebook at the scene of the incident or based upon something which the officer saw at the scene and can clearly describe.

This may sound to be a daunting situation but it is not nearly so bad as it would appear to be and it all boils down to the need for a security officer to be a quietly efficient person able to notice things without any great show of doing so and to be capable of making lucid written notes of the things he notices and of what he hears people say. Then, with the aid of his notes, his description of what he saw and heard and of the action taken by him or others at the scene will be consistently the same accurate story no matter how many times he is asked to repeat it nor how much he is questioned about it. A story told in court under oath or before an official tribunal or board is called evidence but that fact should not change the substance of what the witness has to say – it will only change the setting in which the story is told.

In criminal cases the burden of proof rests upon the prosecution to establish by evidence that the accused person has committed the offence of which he is accused. When a court is satisfied that the prosecution's case requires an answer or explanation from the accused then the accused must present his defence and call his witnesses and/or give evidence on his own behalf.

Evidence is said to be of three kinds – direct, circumstantial and hearsay. *Direct evidence* is that which directly proves the point at issue, for example – who committed a crime or who behaved in such a way that an accident was caused. Direct evidence is the evidence of an eye-witness who saw the crime being committed or who saw some misbehaviour which resulted in an injury to someone.

Circumstantial evidence forms by far the greater volume of evidence and is that which proves attendant and surrounding facts of an incident of which there was no eye-witness. Circumstantial evidence can be so strong that there can be no doubt as to who committed a crime or who was responsible for an accident. Circumstantial evidence can also be used to corroborate the story of an eye-witness.

Evidence is said to be *hearsay* and not admissible as good evidence when a witness speaks of something he has learned or heard from some other person and does not know for himself. There are a few exceptions allowed in the laws of evidence whereby hearsay evidence is admissible. One such exception is that a witness is sometimes permitted to give evidence of what he heard said by a third party provided that he can also say that that statement was spoken in the presence and hearing of the person who is the defendant in the case under review. The purpose of giving such evidence is in order to be able to also give to the court an expression of the defendant's reaction to hearing that statement at the time particularly where this can be said to be an admission of responsibility or guilt for the offence charged. The person attributed to have made the statement will also be called as a witness to tell his own story and the hearsay evidence given by the first witness will be of value as corroboration of that story. Other exceptions concerning hearsay evidence refer to dying declarations and the evidence of children in assault cases or of immediate complaint by a woman who has been the victim of rape.

Giving evidence is the simple action of narrating facts which have been noticed by the witness using normal senses of sight, hearing, smell, taste or touch. A security officer who is called upon to report what he knows should follow the same pattern and should narrate factual evidence of what he heard, saw, touched, smelled or tasted without becoming involved in expressions of opinion.

If he heard a male or female voice shout 'Help' he should say just that and should not evade the point by saying that he heard what sounded like a man or woman calling for help. Expressed in this way the evidence lends itself to challenge that the witness is not sure of what he heard. If the witness saw Smith, Brown and Robinson walking towards him he should say so and should not become involved in saying that he saw three men walking towards him whom he now knows to be Smith, Brown and Robinson because this latter form of evidence leaves room for doubt as to whether or not Smith was one of those three men.

If the witness touched a bottle containing fluid which he tasted and found to be whisky he should say just that and should not involve himself in a colourful description of the bottle and contents which tasted like 'distilled and spiritous liquor of an intoxicating character and prepared from malted barley'. If his taste tells him it is whisky he should say so and that is one expression of opinion which is most likely to be acceptable. Those illustrations apart evidence of opinion is usually reserved for expert witnesses with specialized knowledge such as medical men, psychiatrists, geologists, and acknowledged experts in handwriting or fingerprints.

A witness should speak for himself and should leave others to speak for themselves whether or not he knows what they have to say. Some of his evidence will be in support of or corroboration of what others have to say or alternatively others may corroborate what he is saying. There will be yet further instances in which others will contradict what he has to say or will allege that he is mistaken in his evidence of what he saw or heard or that he imagined what he says that he saw or heard. The witness must not be put off by such assertions and must stand by what he knows to have happened regardless of persuasion or inducement to change his mind.

If a witness shows during his evidence that he is honest and sincere in his endeavour to be fair and frank he will be doing what is right and should not be fearful of those accusations or challenges which are intended to test his sincerity or to undermine his confidence in himself. The best witness is one who can tell his story in the natural sequence of events as he remembers them and spoken in ordinary language without embellishment, dramatization or prejudice. The witness may be assisted by a barrister or solicitor to keep his story in sequence or in giving a clear explanation of facts known to him and that is known as an examination-in-chief.

The witness may then be asked questions by a legal representative of a defendant or by the defendant himself if he is not represented and that is known as cross-examination.

A witness being cross-examined should understand that the questions put to him are intended to test out his evidence for accuracy and not to make him look foolish or to prove him to be hostile or biased and he should not show those tendencies. The witness should carefully weigh each question so that he gives a polite considered answer limited to as few words as possible. If the short answer to a question is 'Yes' or 'No' that is all that should be said. If the person conducting the cross-examination requires clarification of that answer he must phrase fresh questions. Cross-examination is not the time for a witness to expand upon his evidence-in-chief. If a witness is asked a question to which he truthfully does not know the answer he should promptly say so. If a question appears to be susceptible only to a misleading answer it should be answered straightforwardly and the explanation added afterwards. If a witness feels he is being subjected to unfair questioning he may address the court directly: 'May I explain M'Lord?' The court will protect a witness from an unfair attack under cross-examination so long as the court is satisfied that the witness is being truthful and sincere. Should cross-examination produce from a witness the possibility that he has further evidence to give the first representative may re-examine him on those matters for the information of the court or board of enquiry.

Evidence by Written Statements – Criminal Justice Act 1967 Section 1

Attention is called to Chapter 14 with regard to the conditions under which evidence may be entered in the form of written statements so that witnesses may be excused from attendance at committal proceedings.

Perjury

By the Perjury Act 1911 it is an offence for any person lawfully sworn as a witness or interpreter in judicial proceedings to wilfully make a statement material to those proceedings which he knows to be false or does not believe to be true.

Judicial proceedings includes any hearing by a court or tribunal where the witnesses testify under oath. A statement material to the proceedings is one which might affect the decision reached at, or the outcome of, the proceedings.

Under the Perjury Act it is also an offence to make a false statement on oath or otherwise to obtain a formal registration or certificate of marriage, birth or death or to sign a statutory declaration which contains false or untruthful statements.

Security Officers Called as Witnesses

The following golden rules should be observed by security officers called as witnesses:

1 Be neat and tidy in appearance. The question of whether uniform or ordinary clothing should be worn is a matter for guidance from

the Security Manager but it would be quite proper for a security officer to attend court in uniform if the circumstances requiring his attendance at court occurred whilst he was wearing a uniform. It may assist the court for the witness to present the same appearance in court as at the scene of the incident under discussion.

2 Be punctual and allow time for delays on a journey in order to arrive on time.

3 Take with you your notebook or any document or article upon which you made original notes at the scene or immediately afterwards. Read over those notes to refresh your memory before you are called to give evidence and have them ready to refer to should you feel the need to do so in the interests of accuracy during your evidence. Do not read from your notes otherwise.

4 If before you are called to give evidence you are handed a copy of a written statement which was taken down at your dictation during an investigation into the incident you may read over the statement to refresh your memory as to detail and sequence of events but you may not take that copy into court nor in any way refer to it whilst giving evidence.

5 If your evidence is to be given before a judicial hearing where you will be asked to take the oath do read words that are on the card handed to you by the Court Usher.

6 Address a Judge in a High Court or Crown Court as 'My Lord'. Address a Judge in a County Court as 'Your Honour'. Address a Magistrate as 'Your Worship' and address all others in a court of law as 'Sir'. Address the Chairman of any panel or board and all legal representatives as 'Sir'.

7 Stand erect when giving evidence and address all of your evidence and answers to questions to the Judge, Magistrate or Chairman in a clear voice.

8 When giving evidence never answer a question by asking one and never become involved in an argument with the bench or bar or with the defendant. Once you have given an answer to a question do not be provoked into an argument because you are asked a question for a second time or because your answer is challenged.

9 During any breaks or adjournments in proceedings do not enter into conversation about those proceedings with anyone other than the legal representative of the party for whom you are a witness.

10 If you have incurred expenses in travelling to the court or enquiry

you should report the fact to the representative who has called you as a witness or to an official of the court or place of enquiry upon your arrival.

Many people feel quite deflated and ill at the prospect of being asked to give evidence because they have a fear of saying something that will cause hurt to some other person or of saying something which they may later have to admit was mistaken and wrong. To the trained security officer neither of those fears should be relevant because (a) if what has to be said is harmful to some individual it will only be due to that person's misdeeds or lack of self discipline and thereby the harm will have been brought upon himself and (b) if the officer's sense of fairness and honesty prevails he should never find himself in the position of retracting his evidence because it was mistaken or wrongful.

All evidence should be looked upon as a mere expression of the truth and as such there are no pitfalls or traps which should worry a security officer into refraining to act when action seems to be the right and proper course to take.

PART IV

Guidance for Security Patrolmen

Introduction

There is no reason whatever why security patrolmen should not set their feet upon the ladder of promotion by making a study of any or all of the matters dealt with in Part III of this book and indeed they should be encouraged by their seniors and by company management to do so. Schemes which provide bonus payments or higher salaries for those who produce evidence of academic achievement are fairly commonplace in the business world and those schemes should recognize acceptance as a Graduate of the Institution of Industrial Security which is within the reach of any rank and file security man who applies his mind to a little formal study and written work.

The chapters contained in Part IV are intended to give guidance towards the more efficient discharge of security duties and to give to those duties a meaning and purpose. It is hoped that a study of these chapters will create an appetite for a fuller understanding of the work so that new security recruits and those who have no previous training will take an interest in the content of Part III of this book.

CHAPTER 21

Fire Prevention Measures

All large fires from small fires grow and the object of fire patrolling is to seek out incipient fire to prevent it from smouldering and developing beyond that stage.

One of the most common causes of fire in industry is from overheated power operated friction plant or electric motors and although it is the task of machine operators to guard against such overheating there must arise occasions or locations when or where that supervision is not being given at the critical time hence the extreme value of the watchful eye of a patrolling security man. The observant patrol will also take prompt action to reduce any overloading of an electric power socket and will notice electric cabling which is perpetually moving so as to wear away the insulation; or is continually being rubbed by the movement of plant, or a door, or a piece of furniture with the same wearing effect; or he will recognize the damaging effect upon cable insulation of continuously applied heat as when a cable is laid across a hot pipe or radiator; or will see the potential danger in loose cabling being laid across or tied to water piping or to a water tap just to keep it out of the way. Women workers in particular will not realize the hazards of such matters and their working areas must come under special scrutiny from security patrols at regular intervals to eradicate any such dangerous innovations that have taken place in those areas.

Another major cause of fire is the misuse of space heating appliances because employees will use them for quick drying of wet clothing, or will turn the power higher than is reasonable for the environment and thus produce tinder dryness in the compartment and its contents; or will place covers over ventilators or air circulation vents so as to reduce draughts. Any of these actions or others of a similar character must be promptly reported by security so that they do not become established features of a working location.

High in the list of fire causes must of course be those caused by smoking materials being carelessly discarded. The reckless disregard for safety of premises, property and persons which is displayed by a great number of persons who regularly smoke has to be seen to be believed. Fewer fires may now be attributed to smoking than say two years ago but this is because the number of smokers has considerably reduced and not because of any greater caution being exercised by those who do smoke. Pipe smokers will knock out hot embers into waste paper baskets (which should have long since been replaced by metal bins); lighted matches will be discarded or 'flicked' in any direction and upon any floor covering with no attempt to put out the flame or to allow time for the matchhead to cool; partly stubbed out cigarette and cigar ends will be allowed to smoulder in ashtrays (or on dinner plates) and will then fall or be tipped onto furniture or paper-filled rubbish containers; the principal offenders are of course those whose addiction forces them into clandestine smoking in places where, and at times when, smoking is prohibited – one hastily lit half cigarette – one carelessly discarded match or one smouldering stub in an area surrounded by polystyrene and the resultant fire could cause a fire loss claim upon insurance for hundreds of thousands of pounds and possible loss of employment for several hundred workers for at least a year. Indeed smoking rules and prohibitions do demand stringent enforcement and free lance security patrolling is one essential action towards that enforcement.

No less essential is the vigilance demanded in areas regularly used by the public and for very much the same reasons except that it is noticeable how many more places of entertainment, shops, museums etc are prohibiting smoking by the public on their premises.

Consistency of action persistently carried out with regimental routine will go a long way towards minimizing the risk of fire and towards ensuring compliance with fire legislation. One such summary reads as follows:

Daily at Start of Business – ensure that –

1 Doors which are designed as escape routes are unlocked and that escape routes are unobstructed.

2 Fire and smoke doors are swinging shut and that cleaners have not wedged them open.

3 Fire detection and alarm systems are tested at a specified time (unless, as in some systems, testing is required at less frequent intervals).

Daily at Close Down – ensure that –

4 *Fire Doors and Fire Shutters are Closed* This is an essential condition both for insurance purposes and to reduce the loss risk.

5 All outside doors, windows and other means of access are in a good state of repair and are secured against intruders who may cause fire to cover up other misdeeds.

6 Furnaces and boilers are, where appropriate, safely out.

7 Heating apparatus and mains switches are, where appropriate, turned off and the plugs removed from wall sockets.

8 Inspection of whole premises takes place. Special attention should be paid to store rooms and other areas rarely visited, including areas where maintenance staff or contractors' employees have been working, to check against any incipient fires.

Frequently – ensure that –

9 Free access exists to hydrants, extinguishers and fire alarm contact points. Extinguishers are not tucked out of sight in the angles of girders or beneath office desks or under piles of rubbish and old packing or behind stacked goods.

10 Smoking takes place only where permitted; that an adequate number of ash trays are provided; that no smoking takes place during the last half hour of the working day. (This might well be a good rule even in areas where smoking is normally permitted.)

11 Naked lights are not being used (unless specially authorized eg during a power failure).

12 Combustible materials are kept well clear of all heaters, including steam pipes.

13 No unauthorized heaters, gas rings, electrical wiring or fittings are being used and that wall points are not overloaded whilst others in less accessible positions are unused.

14 Combustible materials are kept clear of light fittings.

15 Glue kettles, crucibles, pressing and soldering irons or similar appliances are clear of combustible materials and are provided with non-combustible stands or holders.

16 Flammable liquid containers are closed and are clear of any sources of ignition (preferably placed in an outside storage so that they do not become an additional hazard for fire fighters).

17 Waste bins and ashtrays are emptied at regular intervals, and without fail at the end of the working day; that there are no unnecessary accumulations of combustible process waste, packing materials or floor sweepings; that safe disposal or incineration of waste takes place. Sacks of waste paper should not be stored overnight in lifts or corridors of office blocks.

18 Drip trays are emptied.

19 Workmen's clothes and overalls are kept in appropriate specified places away from combustible materials and sources of heat.

20 Electric motors are kept clear of all accumulated materials.

21 Gangways are kept unobstructed.

22 Special care is exercised by maintenance staff and contractors' employees when operating cutting and welding equipment.

23 Specified escape routes are kept free from obstruction.

24 Re-arrangement of plant or furniture or new plant has not blocked access to power points, fire call points or fire extinguishers.

Weekly – ensure that –

25 Fire extinguishers and other fire appliances are charged and that fire buckets are filled with sand or water. There is general neglect in this direction, the buckets become the repository for rubbish and scraps of all kinds which may be fuel to a fire.

26 Fire instructions, fire exit and no smoking notices are clearly displayed.

27 Alarm and/or sprinkler systems are tested.

28 Sprinkler and fire detector heads are unobstructed by stacked goods or structural modifications.

29 Goods, particularly those on the factory floor, are neatly stored so as not to impeded fire fighting.

30 Clear spaces exist around stacks of stored materials.

31 No non-essential storage takes place in workshop areas or in corridors, gangways, staircase landings or similar locations.

32 Flammable liquids and other hazardous goods in workshops are checked to ensure that stocks are kept to a minimum and stored in unbreakable identifiable containers.

Periodically (as appropriate) – ensure that –

33 Cupboards, lift shafts, spaces under benches, gratings, conveyor belts and spaces behind radiators are kept free of rubbish and dust.

34 Girders and ledges are free from dust.

35 Weeds and grass cuttings are removed from the vicinity of buildings.

36 The water supply is examined and tested for pressure quarterly.

37 Earth leads are connected and that worn cables, broken plugs, overloading and other defects in electrical equipment are notified.

38 Floor drains and scuppers are not blocked.

39 Roof leaks are repaired.

40 Gates and fences which might provide easy access for children or other potential intruders are made secure.

41 Routine inspection of fire extinguishers takes place and is recorded.

42 Maintenance of any special extinguishing systems eg dry powder, carbon dioxide, takes place.

43 A check of lightning conductors takes place.

44 Security procedures for visitors are being observed.

For effective fire patrolling it is advisable for a security force to compile its own list of fire hazards peculiar to its areas of patrol and whilst that list will include many generalities such as unplugging office equipment, closing fire break doors, clearing obstructions from fire points and fire call points and the switching off of unwanted sources of heat or ignition there will be many other individual hazards which only experience on site will recognize as matters to which the attention of each and every patrolling security guard should be given.

Where fire contingency planning has been carried out and where there is regular and efficient fire patrolling the possibilities of fire occurring, or of developing if there is an outbreak, are reduced to a minimum. The less planning there is or the more tardily the premises are patrolled, the greater are those possibilities until no planning and no patrolling provide situations where fire risks are extremely high to both premises and personnel. Fires do not happen – they are caused by carelessness, by neglect, by thoughtlessness, by irresponsibility, by wanton misbehaviour or out of malice.

People at work are a mixture of human moods, depressions, repressions, hatreds, loves, anxieties and fears which cause them to be forgetful, careless, reckless, wilful or spiteful so that they create fire situations or leave behind smouldering hazards from which fire will eventually develop. It is to allay the results of those human failings or vices that vigilance and preparation are essential. Add to those foundations for fire such matters as the work of criminal intruders or mischievous trespassers who regard the making of fire as just one more cheap satisfaction for their ego and the importance of vigilance and preparation is inescapable.

All scenes of fire must be examined for cause and if there is the slightest hint of evidence that some person may be blamed and can be identified a full investigation should be mounted with no attempt on the part of anyone to overlook or cover up what has occurred. The action of security should follow the lines of investigation as set down in Chapter 18 and if police are to be called to an alleged arson there should be as little interference with the scene as possible after the fire fighting has subsided and until police arrival.

It is regrettably a failing amonst some security officers that they take the view that because satisfactory arrangements for fire fighting, for raising an alarm of fire, and for persons to safely escape from buildings, are made and because there is good insurance cover effected without too many arguments with the insurers, very little further action or attention needs to be paid to fire matters. Thereafter security activity and interests for those officers are focused upon the crime aspect of the work but if they would stop to consider the long term effects of fire upon a business they can only conclude that those effects could be far more disastrous than all the accummulative effects of the crime which they prevent or detect. Fire prevention therefore ought to receive the greater share of security time and interest during normal working hours and at all other times. Security's responsibility in this respect has to some extent been stepped up in recent times by their position under the law for supporting management to provide and maintain a safe working environment – shortcomings in this respect could form the grounds for critical comment from works safety committees.

It was said in Chapter 13 that occupiers of premises used as places of work may in the future meet closer scrutiny of their fire precautions and fire prevention arrangements in compliance with conditions of fire certification. It is better for companies with the support of their security departments to go into the inevitable voluntarily and in good form than to be dragged in with disgrace!

For future reading on this subject see –

Planning Programme for the Prevention and Control of Fire
Planning Fire Safety in Industry and Commerce
and numerous other advisory books and leaflets published by The Fire Protection Association Information and Publications Centre, Aldermary House, Queen Street, London EC4N 1TJ
Fire Fighting in Factories Part 10 of the Health and Welfare series published by HMSO Government Bookshop
Fire Precautions Handbook by G W Underwood published by Gower Press Ltd, Epping
Guides to the Fire Precautions Act 1971. No 1 Hotels and Boarding Houses published by HMSO Government Bookshop.

CHAPTER 22

Security Patrolling – Some Practical Points of View

Security patrolling is carried out in business premises the world over and has been going on for generations under the guise of fire watching, night watching, caretaking or some such similar title. Today it is more sophisticated because security work has become more complex but the aims are the same namely to ensure that the premises will be open for the return of the next shift of workers and to keep marauders and itinerant thieves from reducing the profit margin of the business.

Radio Equipment

Today the patrolman has the advantage of many security aids not available to his watchman forbears the most valuable of those aids being the two-way speech R/T the walkie talkie set. The benefit to be derived from such instruments may be summarized as follows:

1 The facility which it provides for a patrolman to maintain contact with his colleagues in all site locations and under all conditions. He no longer is alone on patrol and therefore has courage and a feeling of security to go into situations which he would not attempt but for his R/T.

2 The facility which it provides for a patrolman to call up assistance without leaving the spot upon which he makes that decision. This may be a fire or accident incident or an observation or confrontation with intruders and/or suspects. It could be an incident in which he is already injured or in personal danger, or it could be that he has come upon an unusual object which demands investigation immediately as a possible explosive device.

3 The facility which it provides for a patrolman to seek advice and guidance from a more experienced colleague or from his supervisor or manager. It is so much more efficient to take instruction over an R/T set for a particular line of action to be followed or to comply with an order and then to continue patrolling than to expend time upon a long trail back to the security office to obtain those instructions or orders with the possibility of no continuance of the patrol after compliance.

4 The facility which it provides for a supervisor or manager to direct the movements of his men on patrol to meet changing situations and to deal with incidents.

Just as it is unthinkable nowadays for a police force or fire brigade to attempt their work without the communication facility of R/T so it is unthinkable for security to attempt similar work without that aid. Call up bleeping systems, coded messages over public address systems, scheduled ringing in arrangements by telephone are all antiquated, backward, and outmoded as compared with two-way speech R/T and except on the smallest of sites no security department should be asked to function without this modern aid which transforms a collection of disjointed units into a co-ordinated team.

Another important aid to patrolling is a pocket whistle which in many situations will substitute for an R/T which has become inoperative or ineffective. It may be worthwhile narrating the experience of one guard who left his walkie talkie in his patrol van for a few seconds whilst he examined the cause of an unusual light in a locked building. The van was driven off by intruders leaving him with no access to a telephone or other means of communication except his lungs and a whistle by which means he was able to raise an alarm and to cause the van to be stopped at the exit gates.

Guard Dogs

Some people will say that security patrolling is not effective unless carried out by a dog handler with a trained dog and doubtless there are numerous locations and circumstances in which that is true but there are also many more locations and circumstances when dog handlers and dogs are not only impracticable but are not economically possible. It is the rule that one dog can only properly respond to one handler so the sharing of security duties between all members of a security shift is immediately unbalanced if one member of the shift is a dog handler doing all the patrolling whilst others are tied to watchroom duties. It would indeed be expensive to employ a dog handler upon office routines.

There is much advice and instruction in print about the advantages and disadvantages of employing dog patrols and security opinion is usually fairly clearly divided between those who favour dogs and those who do not. It is a regrettable fact that one group seldom sees or acknowledges the points of view held by the other and when an essay or journal article is written about dogs in security there is frequently a one-sided opinion expressed which detracts from the value of the essay or article as a discussion or as a basis for discussion upon this controversial topic. Suffice it here to draw the readers particular attention to the contents of the Animals Act 1971 which updated the law with regard to

1 damage caused by animals of a ferocious or dangerous nature; and

2 damage or injury caused by animals not normally regarded to be dangerous but likely to be vicious if not controlled.

Guard dogs come under category (2) above and the Act is very specific about where liability shall lie in the event of damage or injury caused by an uncontrolled guard dog. The Home Office added weight to the law by publishing in January 1975 a *Code of Practice for the Use of Guard Dogs* which reads as follows:

1 All persons and organizations who provide dogs for security purposes, for hire or reward, should

   i keep a register readily available of all dogs in which details of each dog will be recorded;
   ii keep a log-book of all hirings, which should include the names of the dogs and handlers; and
   iii be adequately insured against all claims.

2 No dog should be used for security purposes unless it is fully and properly trained to such a standard that it can be kept under adequate control at all times.

3 Dogs should be kept in a healthy condition, properly kennelled, fed and watered.

4 Every dog should be used under close supervision at all times (preferably accompanied by a suitably trained handler.)

5 Unless under the immediate control of the handler, dogs should be used only on premises or in areas that are reasonably proof against escape and unauthorized entry.

6 Dogs should be transported only in vehicles which afford adequate protection against escape by, or injury to, the dog. The dog should be in a compartment separated from the driver of the vehicle.

7 Warning notices should be displayed where dogs are being used for the purpose of guarding premises.

This Code of Practice was followed by Section 1 of the Guard Dogs Act 1975 which became effective by the Secretary of State's order on 1st February 1976. Section 1 Guard Dogs Act 1975 provides –

1 A person shall not use or permit the use of a guard dog at any premises unless a person ('the handler') who is capable of controlling the dog is present on the premises and the dog is under the control of the handler at all times while it is being so used except

while it is secured so that it is not at liberty to go freely about the premises.

2 The handler of a guard dog shall keep the dog under his control at all times while it is being used as a guard dog at any premises except –

a) while another handler has control over the dog; *or*
b) while the dog is secured so that it is not at liberty to go freely about the premises.

3 A person shall not use or permit the use of a guard dog at any premises unless a notice containing a warning that a guard dog is present is clearly exhibited at each entrance to the premises.

The penalty for a breach of the Section is a fine not exceeding £400 triable summarily.

The remainder of the Act authorizes the Secretary of State to make regulations permitting local authorities to licence premises used as guard dog kennels and to prescribe conditions attaching to such a licence if they think fit. Those regulations are not implemented at the time of going to print.

Security Keys

Equipped with weatherproof clothing, his walkie talkie set, whistle and maybe his guard dog our patrolling guard will be ready for a period of patrol when he has collected from the security office a master key or several submasters or a bunch of individual keys with which to gain access to buildings to be visited and inspected upon his patrol.

A master key is of course ideal and reduces the weight of his equipment when key suited locks have been fitted. Employers and architects might well bear this point in mind when ordering or designing new premises or when alterations are to be made to older buildings.

In a building where confidential and valuable articles are properly secured in locked cabinets, cupboards or compartments, interior room doors should not be locked and our patrolling security guard needs to carry only one key with which to enter the building. Freedom of movement within an unattended building is desirable for fire fighting, for a quick search for suspected intruders or for an alleged bomb plant.

Interior doors which are closed but not locked provide the necessary fire breaks without the possibility of incurring unnecessary and expensive damage which fire fighters and thieves might cause to prize open doors if they are locked. If it is considered to be essential that some interior doors are locked then those locks should be submastered with the keys available to security.

Clocking Systems

Our patrolling guard may require a watchman's clock for recording his round of check points. The poor features of some clocking systems seems to be the weight of the clock to be carried by an already heavily garmented and equipped guard but there are available systems which operate by insertion of a key at a key point causing an electrical impulse to record the time on a clock tape located at a central position eg the security office. Clocking systems, like dogs, have their advocates and adversaries but overall there is the great advantage that a good clock record is as much protection for the conscientious patrolman as it is a bugbear for his less conscience stricken colleague. Clocking systems should be so worked that they do not give to other employees an indication as to where and when they may next expect the patrolman to be.

In the long term a clocking system usually justifies the cost of installation provided it is given proper maintenance which is necessary from time to time to overcome interferences intended to make the system unworkable.

Patrolling Duties

What our roving guard should look for on patrol is essentially a matter for individual site demands and can be as diverse as between a dripping fuel oil supply line to a broken stay on a fence support; a power supply not disconnected from an office machine to overalls hanging across a hot radiator in a locker room; an overhead lamp not lit, to ice on the steps of an iron fire escape stairway. The list is inexhaustible and must include any matter which has a bearing upon the safety and security of the premises as part of the general security function of protecting the company's assets against loss by theft, fraud, fire, waste, damage and trespass.

Frustration is frequently felt by patrolling officers who report defects only to find that no action is taken to remedy the fault or to repair the damage. The same matter repeatedly comes to notice and there develops a feeling that such reporting is futile and that the object of good patrolling is groundless. Herein lies the task of security supervisors and managers to take follow-up action with a view to causing remedial action by technical and maintenance departments or by office managers or other departmental chiefs and thus ensuring that the morale and interest of his men is maintained.

Security patrolling is one of those peculiar jobs containing a great variety of interest and yet inducive to boredom. It is a job which causes time to fly or to hang insufferably according to one's mood, degree of interest, energy and zeal. It is a job which when performed at a reasonable rate can be rewarded by a sense of achievement, if performed too leisurely or apathetically can bring boredom, criticism, accusations of inefficiency and lack of enthusiasm but if performed too smartly and zealously will attract complaints of over efficiency, of heavy handedness or even victimization of individuals. The village bobby did not make himself appreciated and respected by exercising high handed authority, by slow wittedness or by apathy, he earned acclaim by adopting a middle of the road – fairness to all – firmness without fighting – attitude towards the villagers and their problems and so it should be with the security patrolman. He may not shrink from firm action when necessary but must abide his time when appropriate; he may not hang in the shadows too long but must make his move with discretion and timed to perfection. Intelligently selecting those matters for action which are important and leaving unheeded those which are of lesser importance is an art exercised by only those men who are the best in the business. They will proportionately divide their time and attention between

supervision of car parking and the discreet observations upon parked vehicles.

Keeping fire points free from obstruction and removing wedges from beneath open 'fire break' doors.

Visiting all locations which are fire risk areas and at fire risk times such as when offices have been recently vacated by staff or when areas have lately been used by the general public.

Checking the security of computer suite, cash office, safes, strong-rooms, fireproof cabinets and the like and removing those keys to safety when they have been absentmindedly left dangling in the locks.

Visiting other vulnerable places.

Checking upon the conduct of contractors' men and visiting drivers to ensure not only their own safety but that of all others on site.

Stopping dangerous 'horseplay' with equipment, wheeled trucks and other people's personal property.

Ensuring that there is adequate light where required without wasteful misuse of power supplies.

Seeking information as to what, where and when there is a threat to the security of premises, plant and people.

Examining perimeter fencing and any spare land for evidence of unauthorized access to or agress from the premises and taking proper reporting procedures of such evidence.

Examining the same areas for property concealed and awaiting final removal off site by thieves.

Examining unoccupied buildings for signs of intruders or of unauthorized interference with normal locking up measures.

Examining occupied buildings and storage areas for signs of criminal activity, accident hazard or fire potential.

Giving general security attention to loading bays and parked lorries, to railway sidings and standing rail wagons and to a wharf and wharf buildings if any.

Removing small offcuts and items which present an accident hazard and taking other measures to have potential accident situations cleared up by liaison with departmental supervisors.

Observing that the movements within premises of materials, equipment and finished goods conform to normal patterns of procedure and taking investigatory action when they do not.

Assisting at the scene of an accident with first aid treatment, obtaining medical aid as necessary and observing the location for verification of the alleged cause.

Keeping in mind the company rules concerning smoking, betting, gambling, tippling, speeding, unruly conduct, private trading and unauthorized clocking in and out of the premises and reporting observed breaches of any of those rules.

Daytime Patrolling

Patrolling premises during business hours has many pleasant rewards because it is then that the patrolman gets to know the business, how its products are made, which are the vulnerable areas and who works in which department. There is time for a friendly word here, a nod and a smile there, a short consultation with a supervisor and a briefing on site from the security chief. There is time for a man to use his personality, to create confidence in himself, to invite confidence from others, and to 'sell' security as a service to the company and its employees.

There is time to take a look at the fire appliances and the first aid boxes to see whether they have been used and replaced empty or half empty to their normal resting place. Each member of the department could be allotted the appliances in a particular building or section of the site as his special responsibility so that each man develops a sense of pride to ensure that when required in emergency, his equipment is not going to be found wanting.

There is time to warn visiting drivers about exceeding the site speed limit and to point out to them that that limit is not a whim of the security department but the wishes of the workforce in the interests of safety. The drivers of vehicles seen to be moving on site with box van rear doors swinging or with tilt bodies raised at an angle should be warned of the dangers they cause to personnel and premises.

Patrolling on or near a wharf or jetty occasionally raises the problem of what action should be taken by security about the smuggling of goods from coasters, barges or other waterborne craft. Smuggled goods are those clandestinely landed to avoid the payment of customs duty levied upon them and include intoxicating spirits and tobaccos. Such landings do not usually affect the security of the site occupier with the exception that he is liable in law if he knowingly permits the premises to be used for that purpose. So, it is a matter for security to ensure that their employers do not become liable for prosecution by the habitual use of the wharf, jetty etc for smuggling.

Investigation could be made to identify persons and property involved in any landing of goods illegally but it is far better for a formal report to go under confidetial cover through top management to HM Officers of Customs and Excise who have full authority in law for dealing with that information. There are small rewards payable to the informants. If the activity concerns the landing of larger quantities than the odd bottle or two or the single box of cigars then it might well be a matter for immediate police investigation pending the arrival of Customs Officers.

Extending a patrol to the employees car park is good security practice because it exhibits a security interest in the welfare of employees and can have beneficial effects when keys left in cars are impounded and a note left on a clock card to that effect, or when a phone call tells a driver that his lights are left on or there is a flat tyre which the driver may get permission to repair during his meal time. Employee car parks are also the occasional target for 'outside' thieves and they should be deterred by regular uniformed guard visits to the park. 'Showing the uniform' also has good effect in the vicinity of contractors' stores and tea huts, in washrooms and cloak and locker rooms, around lorry waiting parks and at any point inside the perimeter fencing at any time. Good patrolling includes watching people moving about on site – in areas where they have no cause to be; in buildings where they have no cause to be; carrying rolled up overalls for no good and clear reason except that they may be a cover up for the illicit removal of some article across the premises; the loosely hanging jacket or outer coat from the shoulders without arms inserted in the armholes may not be because the wearer is too lazy to put it on properly but because there are bottles or other articles inside the sleeves or in the pockets which do not show when carried in that way. Tool boxes are very handy containers of stolen goods and lengths of piping wheeled on a pipe truck have been known to conceal cable. The movements of contractors' vehicles or visiting scrap merchants vehicles on site during lunch period must always be viewed by the patrol with suspicion because it is at that time when working areas and unattended tools are at greater risk of theft. Indeed patrolling during business hours can be full of interest.

Night Patrolling

Patrolling during non business hours is somewhat more relaxed but should be regarded as a more responsible activity because it is at this time that the company's trust in each man's integrity, sense of duty and conscientiousness comes to the fore. Different skills are exercised and in place of evincing interest in the movements of men and materials and in spreading fellow feeling amongst the workforce, the patrolman now works alone in vast empty buildings or 'midst silenced machinery using his senses of sight, hearing and smell to their fullest extent, to seek out possible intruders, dangers or defects. 'We must get more light over that end of the lorry park and we don't need all those lights on in the office block now that the cleaners have gone.' 'Have they gone? I'd better go and take a look.' 'Fancy me talking to myself but really I'm only thinking aloud – it seems to help.' 'Good job I came in here, they've left that fire on again where they have their tea – must leave the chargehand another rocket I suppose.' 'What's that I can hear – running water – another tap on in the lav I'll be bound.' 'Inside this building it's warm but it's a right storm out there.' Button up and face it I suppose.' 'I wonder what that light is over there – looks like a car outside our fence – courting couple I expect but I must take a look – can't take chances – they've done that before to cover up a job.' 'Old Tom was fooled by that one only three weeks ago.' 'I'll get the car number just in case.' 'Hey there, what's the hurry?' 'Well, fancy them driving off just as I was about to get the number.' 'Must tell Control for the record.' 'Mill 2 to Control – over.' 'Mill Control receiving you Mill 2 – over.' I've just disturbed a car parked on the verge outside of our north fence – did not get the number but it's a light coloured Cortina estate, no other description – drove away along the estate road towards Faverley – over.' 'Got all that for the record Harry. Thank you – over and out.'

'What a good view of the sidings there is from up here. We might get one of those CCTV cameras put up here when they arrive – they say they don't worry about a wind like this.' 'Now for that clocking point at the pump house.' Pumps working okay by the sound of them – no reading on the chart, that's good.' 'Better go down to the old lab now – one good sniff there and I'll soon know whether that old trouble is still with us. I'd hate to get "old smelly" out of bed at this hour.' 'Everything's okay around here.' 'Well now, that's about all up this end.' 'Mill Control to Mill 2, are you receiving me? – over.' 'Mill 2 receiving loud and clear – over.' 'Harry will you go down to low level bridge to help Jock Peters with some galvanizing blown from the roof of the old engine house? – over.' 'Mill 2, message received, will go – over.' 'Thank you Mill 2 – over and out.' 'What a boon – what a comfort this R/T really is – what a damned nuisance, I was just off to get a hot cuppa. Oh well, that's patrolling I suppose!'

CHAPTER 23

The Security Office

The Operations Room

Whatever it is called – be it the watchroom, the gatehouse, the security office or the hall porter's lodge it will serve the purpose of being the base at which and from which the security staff on duty will operate. It is usually a vulnerable place because it is the focal point and control centre for the emergency and non-emergency systems of the premises the details of which it would not be politic to mention here but suffice it to say that it is a place which should never be left unattended and demands far more physical security for its occupants and contents than are usually given to it. The best advice that can be offered is that from the time of reading this chapter every manager, security or otherwise, who is responsible for the security of the company premises should take a good hard look at the protection given to his security base to be sure that it is as good as is reasonably practicable. There is always room for improvement and in some cases any improvement will be better than that which now exists. It was probably like that 20 years ago and we know what has happened in that time!

The men who staff the security office must be top line security officers with good training to acquaint them with their powers and authority. They must be schooled into the most efficient ways of using the equipment provided. They must remember that they are 'on show' as the company's first representatives to be met by visitors and must at all times present a businesslike appearance and manner with no flippancy or frippery displayed within the work area of that office. Visitors should be met with good sense, good discipline and goodwill. It is quite evident that not only must the staff be presentable but also that the entrance and office where visitors are received be in 'bandbox' condition.

When answering the telephone the officer should announce that the call is being received in the security office (or watchroom, or lodge etc) and he should give his name. That is an efficient way to answer a telephone and impresses the caller who is saved the time and irksomeness of asking to whom he is speaking.

Great emphasis should be given to the need for and value of full and accurate records of everything that occurs connected with security and those records must be carefully prepared and preserved for a reasonable period thereafter. An accurate time recorder should be available so that documents may be time stamped in order to eliminate argument as to when hand recorded times were entered. Good filing systems should be introduced so that the chore of looking back over records is simplified. Some records may be computerized for storage so that the volume of stored paper may be reduced and in order to facilitate the production of periodic summaries of security action or other statistics. Many forward looking plans can only be properly assessed if past records are available and the security department is one which should always be forward looking to keep ahead of problem situations.

The Security Checkpoint

The strategic posting of a security guard at an entrance doorway of a building or gateway of a set of premises or site is for the basic purpose of preventing the movement of goods or persons into or out of the building, premises or site without authority or contrary to law.

Chapter 3 has dealt with company policy in the matter of searching as a routine and by random selection but a few comments upon the practical approach to stopping and searching persons and vehicles will not be amiss here.

Close scrutiny of vehicles and persons *entering* the premises is vital to prevent the introduction of unauthorized explosives or other noxious substances. Should a telephone call purport that a 'bomb' has been planted the likelihood that the call is a hoax will be heightened if a close regular scrutiny is always conducted at the entrances.

Another reason for scrutiny is to prevent company premises from being used for private transactions between employees which not only wastes company time but frequently involves unlawful handling of stolen goods. It is well known that factories and similar places of work are used for the distribution, sale and disposal of, goods stolen elsewhere and 'smuggled' into company premises. Some workers supplement their normal earnings by selling to fellow employees on company premises, in company time and for private profit, goods which they buy as agents for wholesale warehouses. In other locations workers handling attractive goods are persuaded to exchange some of those goods for other attractive items brought in from neighbouring factories. Those and similar transactions can be stopped by a company rule which simply makes it a breach of conditions of employment for employees to privately buy, sell or exchange goods on company premises. With such a rule security have good grounds for checking vehicles and hand luggage brought in and could detain in their office goods in respect of which they suspect a breach of the rule might be committed.

There are occasional honest exchanges of goods between employees who find it convenient to make the exchange at work but those transactions should be confined to the gate or entrance where the goods should remain during working hours in a cloakroom adjacent to the security office and controlled by the security staff. Such a cloakroom would be used for holding domestic shopping which shift working women collect on way to work and for outside purchases made by employees during meal break times or whilst they are off the premises on company business. Where schemes operate for employees to purchase goods produced on the premises those goods could be delivered officially to the security office cloakroom to await collection on pass by the purchaser upon leaving work.

Incoming employees who feel aggrieved by a security refusal for them to convey into the premises some package or article should be asked to leave the item at the security office until permission from a manager or supervisor is notified to security, verbally or in writing, that the item may be admitted. Good managers will not see the point of giving such authority.

Outgoing employees in possession of any article which they have 'acquired' during their working hours should be in possession of a passout signed by a manager authorizing the removal of the item. Such a pass should be prepared in triplicate with one copy retained as a counterfoil, one copy to be handed to security and one copy to be carried off the premises with the goods. This copy should be cancelled by rubber stamp or other means to indicate that the goods have been taken off the premises. A passout need not be required if goods have not passed into the premises beyond the security cloakroom.

No vehicle should enter company premises unless essential to the running of the business and every entry should be covered by documentation in the form of consignment or delivery notes or collection orders or covering letters or by a prior intimation to the security office that the vehicle is expected for a stated purpose. This will frequently mean that when staff make telephoned arrangements for a service vehicle to call or for items to be collected etc they should also telephone or send a memorandum advice to the security office to that effect. The regular calls of carriers should be the subject of a standing order of authority for the vehicles to be admitted on site.

Depending upon the nature of company business the degree of record made by security about vehicle arrivals will vary considerably from mere entry of an index number in a register to procedures which include time stamping documents and for endorsing them with the security office stamp or by the retention of a copy or photocopy of each document presented by drivers.

Searching incoming vehicles is a matter for company policy but whether conducted as a matter of routine for every vehicle or whether there is a casual scan of most vehicles and contents there should be a full check made on a spot check basis to maintain a deterrent control over goods being brought onto the premises unlawfully or without company authority.

The same rules and procedures should apply for outgoing vehicles with more emphasis and frequency of searching being applied than in the case of vehicles entering the site. Nothing should be taken for granted or assumed to be in order and no opportunity should be missed for asking questions about the movement of goods not covered by satisfactory documentation. This is an area of activity in which there can be no room for complacency, timidity or slackness. It is an activity in which all members of the security department must act in harmony, because slackness in checking goods brought in must lead to slackness or disharmony in the action to be taken when those same goods are taken out by different people. Only by sustained effort by all officers will the aim of preventing crime be achieved.

It has been emphasized in Chapter 3 that company policy should at the very least permit spot checking of laden vehicles and a full search of every allegedly empty vehicle leaving the premises. Empty road tankers should be weighed or checked by dipstick at the security check point. Regular drivers will become known but it will be bad security if they are not required to follow all standard procedures upon every visit. It is not good enough for a lorry, van or car to be driven through the factory gates with just an exchange of nods between the driver and the security man on duty there. The first time that a driver is allowed to go through with a nod establishes a pattern which he, the driver, will expect not only from that security man but from any others who may be on duty on future occasions.

Because one driver is seen to get away with it others will expect similar treatment so that there grows up in a very short time a custom whereby all regular drivers entering the premises do not stop but merely nod to the guard on duty. Installing a lift barrier re-establishes the routine of stopping and checking documentation.

Special attention should be given to drivers who merely give the magic word 'canteen' when asked where they are going because there will be those amongst them who are not going there to make a delivery, as the word 'canteen' infers, but to consume a company subsidised meal which they know to be available there and/or to sell ladies tights, photoframes, knitting wool, Christmas cards, stationery or a myriad of other articles to the workers who are taking meal breaks. A challenge for documentation would quickly identify such intruders.

Scrap dealers, used tyre merchants, sack cleaners and dealers and collectors for container clearing houses should never be admitted to premises without the production of written orders requesting them to come. Once admitted they will adopt a variety of ruses to remove good condition, undamaged, useable items additional to or in place of those they are required to remove and a strict security check of their loads should be made on every occasion. If their transactions are dependent upon a weighbridge weight every care needs to be exercised to ensure that that weight is not being 'rigged' in some form or another and for that reason the allegedly 'empty' vehicle entering should be searched for items of extra weight which are intended to be discarded on site before loading commences. Such items have been known to comprise sacks of empty bottles, drums of water, steel sheets lying flat on the vehicle floor or on the roof, old tarpaulins or other rubbish stored on the roof of the driver's cab, snow or rainwater, extra passengers who walk away before gross weighing or large dogs which run off the premises.

In this matter of false weighing it is not to be assumed that this only takes place with the collusion of the weigh clerks. Indeed the reverse is usually the case and drivers resort to the kind of activity outlined above in order to produce a false tare weight and to cheat honest clerks who cannot be corrupted. If a weigh clerk is corruptible he will cheat by registering the weight of one vehicle upon the ticket of another for which he wishes to record a heavier or lighter weight and that action needs careful observation and investigation to identify. Suspicion should fall upon the part loaded van or lorry which has been to market before making a collection at company premises because that driver is likely to have a few hundredweights of good cheap saleable items for disposal to employees but will weigh out as though his tare weight has been unchanged.

Difficulties sometimes occur with regard to vehicles such as those of common carriers which are laden with deliveries for numerous consignees. It is so easy for a manager or other persons in special positions to place amongst outgoing parcels in the company stores one or more parcels addressed to themselves at their home addresses or to some other convenient addressee. These parcels containing stolen goods will usually pass out of the premises past security with no check or query being raised and the system can operate regularly for years without knowledge on the part of the carrier that he is being used to aid the theft of goods. The antidote to this activity is for the introduction of a paper control system which accounts for the contents of every package and which is subject to audit or for spot checks to be made periodically upon packages handed to the carrier.

All searches or checks at main doors or gates must be carried out with scrupulous fairness, tact and courtesy so that a wrongdoer disclosed by the system may take no retaliatory action by way of complaint to undermine the case against him or her.

Chapter 3 contains a mnemonic using the word 'SEARCHING' for points to be observed when carrying out a search procedure and this must be specially known by and complied with by those on security office duties.

If information is received that an employee, or any other person, may be taking off the premises without authority some item which is not bulky enough to be evident such as a garment which will be worn, the action of security should not be to stop that person and ask questions but to invite him/her to go through the routine search procedure and to endeavour by that means to identify the item. If the item is not found the person may be released without comment or, depending upon the strength of the information and upon the value of the article alleged to be involved, management may be consulted for guidance upon calling police. If however the search has been thorough there may be little point in calling police on this occasion.

Bomb Threat

Earlier in this chapter reference was made to the all-too-frequent and regrettable incidence of telephoned messages that a bomb has been planted on business premises. The subject is given special comment in Chapter 10 but the part to be played by the staff of the security office on such occasions merits special mention here.

Good routine gate security has very special value in the prevention of a bomb attack or threat of such an attack and a well trained and supervised security staff will know just what to do in the event of information (hoax or not) being received that a bomb has been planted upon company premises. The security office will probably immediately become the communications centre for the emergency and at the same time some members of the staff will be responsible for the security of the contents of empty buildings.

Added to those activities however security officers should bear in mind that an investigation into the source and originator of a bomb threat call (again hoax or not) will be made and that they may be able to assist that investigation by taking the following action.

1 Take a quick look into the street to note any individual who appears to be interested in the premises. Maybe he wants to assess the effects of his phone call!

2 Consider the possibility of the call emanating from an internal telephone and note any characters who are 'first out of the gate'.

3 Keep a sharp ear for snatches of conversation from those who are being evacuated – some young disrupters are quite boastful to their fellows!

4 Take special care to observe stragglers who are late leaving evacuated buildings and search if circumstances merit that action.

5 During an emergency keep normal business telephone calls to a minimum and ask all callers for their number to be called back later.

6 During the 24 hours following a hoax bomb threat call seek casual conversation with employees at all levels to test the possibilities of the call having originated on site.

General

It is quite beyond the scope of this book to attempt to give guidance upon the great variety of other events, incidents and problems that will occur in and around a security office. Efforts are made in other chapters to touch upon some of the more important or more frequent happenings but every business house will be run differently and no two security offices will handle activities in precisely the same way. Much will depend upon the quantity and efficiency of installed 'security aids' and upon the burden of responsibility and trust which each management places upon the security staff.

In the past there has been far too little training for the exacting duty performed in a security office and many men go along with the established procedures which they have learned from their colleagues, without any real confidence that they are doing the right thing. Frequent talks and instruction of a formal nature from the head of the department will dispel some of their doubts and will instil some confidence but it will only be when those men mix with others or receive lectures from strangers at training courses that they will recognize that other security officers meet the same kind of problems as their own and that those problems are solved or handled in very much the same way as in their own company. They then find full confidence in their work and appreciate that they are part of a profession and not just an isolated unit working under the variable whims of their own management.

CHAPTER 24

The Security Officer's Notebook

Throughout this book references have been made to a security officer making entries in his notebook and yet it is known that only about one third of security personnel are issued with, or carry with them a notebook for use on duty. Security personnel who are ex-police officers or who work under security supervisors who are themsevles ex-police officers will naturally carry a notebook as a tool of the trade because police training has shown to them or to their supervisors the benefits and value of having a notebook for immediate use on duty. Other security people have not had that kind of training and experience and need to be convinced that a notebook is essential to the truly efficient performance of their job and to those ends this chapter is directed.

Employers should provide for their security people good quality notebooks in hardback covers of convenient size to fit a jacket pocket. The pages of the book should be numbered and ruled for manuscript entries. Upon receipt of the books they should be numbered serially and placed in a reasonably secure store from which the issue of each book should be recorded as to date and recipient. Thereafter new books should only be issued in exchange for completed books or upon a report that the completed book is required for further duty reference. Completed books should be held in store for at least three years.

The foregoing advice should give some indication of the importance that is placed upon the need for notebooks and upon the stringency and care with which they should be used but to enlarge upon those matters it may be said that –

1 Security work may at any time be the subject of scrutiny and enquiry which can be answered promptly and efficiently by reference to a written note made by a security officer in his notebook. Conversely the absence of a note which should have been made may confirm or clarify the enquiry.

2 A notebook should be maintained by a security officer as a diary of his daily duties and events in a strict chronological sequence which creates authenticity of the entries.

3 A notebook should be used by a security officer to supplement his memory so that he has a means of making immediate written record

of orders and directions about duty matters relating to the future and later to be transferred to an office diary or other aide-memoire.

4 A notebook should be used by the holder to record all relevant and important facts of an incident or happening during his duty such as the time, names and address or other identifying details of persons, and in particular the spoken words of persons who make complaints to the officer or who make allegations against others or who give explanations of their own or other people's conduct. It is emphasized that there can only be one good and reliable record of the spoken word and that is a record made as the words are spoken or made immediately afterwards.

The passage of time plays tricks with the human memory and an interval of only a few minutes is sufficient for an incorrect recording to be made of a statement or remark unless the person making the record is expert and practised.

5 The use of odd pieces of paper for note making is a dangerous practice, even though a full written report is prepared from them shortly after the note making, for the following reasons:

i the only valid and legally acceptable written record is the original note made at the time of an incident or as soon after it as practicable. Written reports prepared from original notes are not so acceptable;

ii pieces of paper are easily mislaid, can be replaced by others and contain no corroboration of the time sequence as to when the written note was made as in the case of notes made in a notebook;

iii a security officer may be in a location where scrap paper is not to hand and he is forced into making notes on the cover of his personal cheque book, or on a cigarette packet or what is more than likely to make no written note at all but to rely upon his memory for note making when he has arrived at his office some time later – in which case the notes appreciably diminish in reliability and value as a true record.

6 A security officer who is called to give evidence in a court of law or before a board of enquiry may find it necessary in order to be accurate in his evidence to refer to his original note of a statement made to him by a party to the proceedings or to a serial number of an article and there are likely to be uncomplimentary comments made by the court or the board about an officer who refers to scraps of paper or other unofficial and incongruous tablets upon which his notes are written.

At court a notebook is not in itself evidence and should not be referred to whilst giving evidence except to be precise about a

> conversation or other detail of information which the witness could not be expected to memorize. In those circumstances the notebook is a document from which the witness can refresh his memory but only if the notes in the book to which he refers were made contemporaneously or as soon as reasonably practicable after the event. When a witness does refer to his notes when giving evidence it is quite lawful and proper for the court to ask to see the notes and for a legal representative for the other side to also ask to see them. How important it therefore is that original notes should have been recorded in a presentable and well kept notebook.

Having it is hoped made out a case for all security personnel to be issued as a tool of their trade with a notebook, it is now necessary to convince those to whom the books are issued of the importance of strictly keeping the book in good condition with entries made in legible writing and reserved only for matters affecting their duty. A notebook is not a scribbling block – its uses have already been referred to and it should be carried at all times when on duty.

The purpose of making daily entries of the date and duty assignment is to prove that any notes recorded during a tour of duty were in fact original notes made during that tour and have not been written into the book as a copy from some other notes on scraps of paper at some later occasion when a need for notes became apparent.

A notebook does not replace an office occurrence book or day book and should not be used solely for writing up full reports from notes made elsewhere on pieces of paper. To do that would be to defeat the whole object of the book which is to provide a place for immediate recording of names and addresses, numbers, times, distances, descriptions of a location, descriptions of vehicles, damage, injury, and verbatim statements or explanations made by persons involved in incidents. Such information recorded into the book as it comes to hand forms the original notes of an occurrence and it is then for local instructions to say whether a further full report based upon those notes shall be written into the notebook or shall be written into an office day book or occurrence book or otherwise for the information of the head of the security department and other management. In those cases when original notes are not followed by a full report in the notebook a brief entry should follow the notes such as 'Full report entered upon Accident Report Form Number ....' or 'Full report entered upon Fire Report Number ....' or 'Full report entered on folio .... of the Day Book' or otherwise as appropriate.

Many years experience in the field of law enforcement and in matters of enquiry and investigation to ascertain the truth of disputed allegations or assertions has shown how important and valuable can be a good written original note which at the time of its making appeared to be insignificant but nevertheless was recorded as a matter of routine. It is therefore not necessary that a security officer should try to decide which matters are of sufficient importance for notes to be made – he should use his notebook for recording factual detail at all times and regardless of whether or not those facts will subsequently be included in a full report.

It may even be that an officer will write in his book an index number of a vehicle and the time at which it was seen to stop outside of the premises just for his own future reference because he has some suspicion about the activities of the driver of that particular vehicle. The officer may never again refer to that note but how valuable will be that note if in one month's time some more tangible evidence of the driver's misconduct is forthcoming and the information is then required for a report or for evidence that the vehicle was seen at a specified time on the date of the original notebook record. It is quite certain that the officer's memory could not have retained that precise date and time for that period and it is most probable that a note on a scrap of paper would have been lost during the month interval.

Another example of the value of notemaking in a book to which ready reference can be made would be when a security officer notices a delivery vehicle being driven on a factory road far in excess of the 10 mph speed limit. The officer may not be in a position to stop the vehicle before it leaves the premises but he does know or can refer to records for the vehicle index number which he records in his notebook together with the time and his estimate of the vehicle speed. When that driver next visits the premises the officer will be able to produce his notebook and say to the driver 'On Friday 12th of this month at 14.30 hours you drove along this road at a speed which I estimated to be at least 20 miles per hour which is double the maximum permitted speed on this site. Will you please make sure that you keep to the limit in future and I am recording this as a warning.' That statement leaves no room for argument and is so much more efficient than for the officer to say, 'About a week ago in the afternoon you drove along this road too fast will you please make sure etc . . .'. The driver's immediate reaction could be 'I didn't come in here last week' or 'I came in three times on one day, what time was that?' Such questions could develop into an argument and that opportunity to challenge what is being said reduces the dignity of the occasion and reflects inefficiency. Using the notebook to confront people with facts which they know they cannot dispute places a security officer in an advantageous position and he will be respected for it.

Great care should be exercised in recording names of persons and the names of roads and districts in order to ensure accuracy. The spelling of a name should always be asked for because Wyke can sound like White, Cocks can be pronounced Coe, Pitman can be spelt Pittman and Wallace sounds exactly the same as Wallis. All names and addresses should be entered in notes in block letters and if the name of a house is given in an address the name of the house should be marked by quotation marks to distinguish it from the surname of the person whose address it is. It is also useful to ask for and to record telephone numbers or private extension numbers to save time if future contact by telephone with a person is likely.

In the matter of recording the spoken word, this should be a verbatim record of what is said. For example a witness may say 'I saw a flicker of light in the hut so I opened the door and then the whole place burst into flames'. That statement recorded in the words of the speaker contains its own explanation of why the witness opened the door of the hut and establishes that there was fire in the hut *before* the witness opened the door. The statement might have been incorrectly recorded in another way, for example '. . . said that when he opened the door of the hut the place burst into flames'. In that form the statement is said to be in the third person and it not only omits an important part of the actual statement but raises in the mind of any subsequent reader of the note, or of a report containing the statement in that form, questions as to why the witness had opened the door of the hut and whether the action of opening the door had caused the interior to burst into flames. This may be followed by an unnecessary waste of time to clear up whether the witness had a lawful reason to go to the hut and whether the fire resulted from some booby trap or prepared situation to deliberately cause fire. Such questions would not arise if the note making had been a verbatim record.

That simple illustration can reflect innumerable situations which may arise in the course of security work where dead-end enquiries are started because a remark or assertion or complaint, or explanation has not been correctly and accurately recorded as a statement in the first person by registering on paper the exact words of the speaker and in quotation marks to indicate that it is a verbatim record.

A verbatim record of conversation is defined in the Oxford dictionary as 'word for word – in the exact words' and it is the ability to record what is said in that way which security officers should endeavour to acquire for their notemaking and reporting procedures and for use in evidence or verbal explanations of events.

So many people adopt a manner of speech and writing which follows the third person principle all of the time and you will hear them say, 'I asked him if he had touched the bottle and at first he said that he had not but when I asked him how the bottle had been moved from the table to the window ledge he admitted that he had moved it'. That episode described in the first person would be – I said to him 'Have you touched this bottle?', he said, 'No.' I said, 'Who moved the bottle from the table to the window ledge?', he said, 'I did.' I said, 'Why did you do that?', he said, 'Because I did not want you to find it.'

It will be noticed that the second version accurately describes the actual conversation whereas the other is an edited version which, in the editing, might give a wrong impression of the attitude of the person being questioned and this can be of as much importance as the substance of his answers. Another aspect is that the verbatim record of the conversation will be accurately reproduced in reports both verbal and in writing no matter how many times it is required whereas the third person version lends itself to many variations because it is not a strict record of the actual words used and is therefore open to challenge that it truly represents that conversation.

The need for making a note of the time *at the time* cannot be too strongly emphasized and the reader's attention is called to what was said in Chapter 18 on this subject that human memory is fallible and that to be accurate about noting the times of incidents one must cultivate the habit of promptly noting down times as events happen.

Security supervisors should not only comply with all recommendations about carrying and using notebooks but should make regular inspections of the notebooks of subordinate security personnel to ensure that proper entries in chronological sequence and date order are being made – to ensure that there are no erasures or obliterations of the entries and that there are no pages loose, left blank or missing.

Efficiency in the use of notebooks is encouraged if each book is handed in to security management at the end of each tour of duty whether or not it contains a full report and whether or not other reporting procedures have been followed based upon notes recorded in the book. This procedure ensures inspection and makes the books available to be drawn at the beginning of the next tour of duty without the risk of a book being left at an officer's home or otherwise being mislaid.

A full notebook does not necessarily reflect a proficient security officer but a well kept book most certainly does and should be given full regard when annual reviews are being prepared.

Conclusion

This book has been written against a backcloth of daily news filled with emergency situations and disasters, of bomb plants at hotels, restaurants and departmental stores and in or beneath private cars parked in the streets; of the findings of large caches of explosives and materials for bomb making; of the seizure and detention of important business men and of wealthy young heirs to family fortunes by politically motivated individuals or by reactionary groups or by gangs of common criminals such as those involved in the Spaghetti House seige in London. The hijacking of planes and trains and of the occupants and casual callers at Embassies seems to have become part of the world-wide scene so that the cold blooded shooting upon his own doorstep of Ross McWhirter in the very presence of his wife passes as yet another incident instead of the appalling tragedy which it is. These activities of political upstarts, reactionaries and common thieves are a clear indication of the future trends in crime which will throw an ever increasing burden upon the forces of law and order who must devote much of their time and resources to the prevention and detection of such violent matters.

Society also must take preventive action against these matters but even more importantly must defend itself against everyday crime which the overstretched law enforcement agencies may not be able to handle.

Home Office statistics tell us that the crime figures are running at eight to ten per cent increase for each half year over similar periods of preceding years and that police manpower tends to be less adequate year by year. Fire statistics inform us that escalating fire losses must result in escalating insurance premiums and ever mounting legislation towards fire prevention.

Society's plan is to enhance the work of police and fire authorities by privately organized security services manned by people willing and anxious to maintain an even balance between right and wrong, between safety and disaster and with the protection of life and property as their perpetual aim.

The book has looked at those people and at the job which they do, has considered the attitudes and policies of management in industry, has outlined the problems and demands of the law and has looked at the means by which industrial management and the security profession can uphold the law and give support to the professional services.

It has analysed the need for, and demand for, training in the fields of crime and fire prevention and if in the future it contributes in any way towards the education and training of those engaged in the profession it will be doing the job for which it is intended. Improved education can only result in improved efficiency which will earn, it is hoped, the respect and appreciation of those splendid services with which the security profession strives to co-operate – whose authority it does not challenge – and for whom the profession holds the highest regard – the British police and fire services.

Index

Parliamentary Acts are grouped under the heading 'Acts of Parliament.' The letter-by-letter system has been adopted.